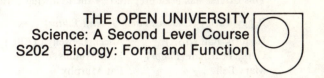

THE OPEN UNIVERSITY
Science: A Second Level Course
S202   Biology: Form and Function

Animal Physiology 1

unit 16
Communication: Nerves and Hormones

unit 17
Blood Sugar Regulation

unit 18
Control Mechanisms in Reproduction.

Prepared by the S202 Course Team

The S202 Course Team

This Course has been prepared by the following team:

Peggy Varley (*Chairman and General Editor*)

Hendrik Ball (*BBC*)	Aileen Llewellyn (*BBC*)
Gerry Bearman (*Editor*)	Pam Mullins
Mary Bell	Pat Murphy
Eric Bowers	Seán Murphy
Bob Burgoyne	Pam Owen (*Illustrator*)
Ian Calvert	Phil Parker (*Course Coordinator*)
Norman Cohen	Ros Porter (*Designer*)
Peter Cole (*BBC*)	Rob Ransom
Bob Cordell	Irene Ridge
Baz East (*Illustrator*)	Steven Rose
Vic Finlayson	Ian Rosenbloom (*BBC*)
Anna Furth	Jacqueline Stewart (*Course Editor*)
Denis Gartside (*BBC*)	Mike Stewart
Lindsay Haddon	Jeff Thomas
Robin Harding	David Tillotson (*Editor*)
Stephen Hurry	Charles Turner
David Kerrison (*BBC*)	Sue Turner

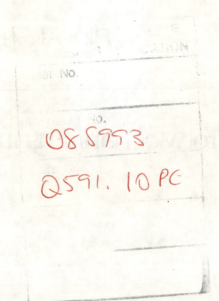
The Open University Press,
Walton Hall, Milton Keynes.

First published 1981.

Designed by the Graphic Design Group of the Open University.

Typeset by Santype International Limited, Salisbury, Wilts, and printed by W. & J. Linney Ltd, Mansfield, Notts.

ISBN 335 16036 0

This text forms part of an Open University course. The complete list of Units in the course is printed at the end of this text.

For general availability of supporting material referred to in this text please write to: Open University Educational Enterprises Limited, 12 Cofferidge Close, Stony Stratford, Milton Keynes, MK11 1BY, Great Britain.

Further information on Open University courses may be obtained from the Admissions Office, The Open University, P.O. Box 48, Walton Hall, Milton Keynes, MK7 6AB.

1.1

Animal Physiology

Physiology is the study of how organisms function. In Units 16–25 we are concerned exclusively with describing the various mechanisms that operate in animals and we discuss the different solutions that enable them to live successfully in their environments. Over the last hundred years or so, biology has moved from the stage of cataloguing and description to that of analysis and experiment, and physiology is a comparatively new discipline. Under the general heading of physiology are included specific fields of interest, such as cell physiology, neurophysiology, functional anatomy, invertebrate physiology and environmental physiology—but one of the important modern developments has been the breaking down of some of the unnatural barriers that have grown up around these specialities. Students of physiology are concerned nowadays with appreciating the major general principles and how these relate to those of other major areas of biology, for example, anatomy, ecology, behaviour and biochemistry.

As you start these Units, bear two points in mind. First, to appreciate how animals function, we must be aware of their external form and internal structure. In this sense, the term 'structure' is broad enough to include the chemical structure of the cell constituents, the ultrastructure of cells as revealed with the electron microscope, the organization of cells into tissues as seen in light microscopy, and the gross anatomy of the animal as revealed by dissection. Clearly, this animal physiology block will build on your knowledge of earlier sections of the Course, and the close relationship between structure and function will be evident throughout. The second point is that an animal's physiology should be related to what we know of its natural environment. It is dangerous to extrapolate too readily our findings in the artificial conditions of the laboratory to the much more complex conditions in the real world. There is a very wide range of natural environments—deserts, cold water and mountains, for example— that different animals have to face. Even in one particular environment, say a hot desert, the ways in which animals cope with the heat are likely to be different, depending on a number of factors such as the animal's size, its diet and its basic organization. All this means that physiologists are often concerned with the variety of physiological adaptations shown by different species of animals in different environments. This is very much the field of *comparative physiologists*, whereas a *general* (or cellular) *physiologist* tends to be more interested in the physiological characteristics that living animals have in common, which are often more obvious at the cellular or biochemical level. These two approaches are complementary and often reflect approaches to common problems at different levels; both views are represented in the Units that follow.

By way of introduction, let us highlight some of the key issues discussed in the Units and describe the logical links between them.

As you learnt in the Science Foundation Course, one important branch of physiology is the regulation and control of physiological processes, and this topic is discussed more fully in Units 16–18. Most biologists consider the nervous system to have the primary coordinating role and, indeed, nerves have the great advantage of conveying information quickly to precise 'target' organs. The first part of Unit 16 introduces some fundamental aspects of the nervous system, in particular the structure and function of the basic unit, the neuron or nerve cell. The long main fibre of the neuron (called the axon) usually conveys information in the form of electrical impulses or changes in membrane potential to the branched terminals of the neuron. There, chemical 'messengers' (called neurotransmitters) are released, and these can directly influence the activity of other nerves or organs—for example by provoking the contraction of skeletal muscles. A second integrating system is the endocrine (or hormonal) system. At one time it was thought that this system relied exclusively on the passage of hormones (which are also chemical messengers) in the blood to distant targets, but we now know that some hormones act over very short distances. Furthermore, there is a marked overlap in function between neurons and endocrine cells and between neurotransmitters and hormones; it is therefore no longer appropriate to regard 'nervous' and hormonal control as distinct and unrelated. Indeed, Unit 16 emphasizes the similarities between them, especially at the biochemical level.

Many regulatory systems maintain stable internal conditions, often in the face of wide fluctuations in external conditions; for example, think of the regulation of human body temperature or the relative constancy of blood pH or blood glucose levels. Many animals have an internal 'steady state' regulated by homeostatic mechanisms. In Unit 17 we consider the homeostatic regulation of blood glucose and illustrate the importance of feedback control, a topic introduced in the Science Foundation Course and in Unit 7, on enzymes. In Unit 17, as elsewhere in the Course, the complexity of the control processes is obvious; a number of important hormones (insulin is the most familiar) can influence the levels of glucose in the blood, and the nervous system can exert an overriding control.

Unit 18 describes how nervous and hormonal systems are involved in mammalian reproduction. Many mammals have breeding cycles, usually correlated with the seasons. Such periodic changes in behaviour and physiological 'preparedness' to breed are regulated by sex hormones such as the oestrogens and androgens; these hormones regulate cyclic changes rather than maintaining a steady state. Animals are able to anticipate the arrival of an appropriate breeding season from various environmental cues. Changes in the length of the day provoke, in some species, a state of physiological readiness to breed; this demonstrates the interaction between the environment and the nervous and hormonal control systems. Two other points of general interest emerge. First, there is great variation in the details of hormonal control, even between different mammalian species: three rodents, the guinea-pig, mouse and rat, though closely related, have breeding cycles that differ in many respects. Second, this is an area of physiology of far-reaching practical application in the fields of contraception and human reproduction. A lot of our concern to understand how animals work comes from simple 'curiosity', but many of the fields you will read about have important immediate medical applications.

Units 16–18 mention a large number of hormones and neurotransmitters and we have therefore included a comprehensive list (Table 1) at the end of the Introduction. You should use it as a summary and refer back to it as you read the Units, especially Units 16–18.

Units 19, 20 and 21 are concerned with the circulatory and respiratory systems of animals, which are closely linked. A circulatory system is a necessity for most animals over a certain size, especially homoiothermic animals (Unit 3), which have a high metabolic rate. Gas exchange between the blood and the environment is often restricted to one site (the gills or the lungs) and the amount of gas (oxygen and carbon dioxide) carried in the blood is increased by the presence of respiratory pigments such as haemoglobin. The movement of large volumes of blood around the body poses acute mechanical problems, and the structure and function of the heart and major vessels are therefore of special interest. Again, there are major differences between species, but some common principles are clear. The control of the blood supply to the respiring tissues is obviously important because in certain stressful conditions, for example during exercise, the amount of blood supplied to the tissues must be greatly increased. Another form of environmental stress occurs when animals find themselves in environments that are lacking in oxygen, and we discuss physiological and biochemical adaptations that allow many animals to tolerate such conditions. Because the physical and chemical properties of different environments are so varied, the problems they pose are quite distinct; consider water and air as respiratory media—water is more dense and viscous than air and contains relatively little oxygen.

This theme of the relationship between an animal's physiology and its environment is continued in Units 22 and 23. Animal tissues are bathed with fluids that may differ in ionic content and water content from the external medium, so salts and water may be gained by the body fluids or lost to the environment. Animals may either tolerate variations in composition of their body fluids or, much more commonly, maintain constant levels of ions and solute in the fluids around and within cells; that is, they demonstrate ionic and osmotic *regulation*. Animals that regulate are more able to cope with a very broad range of aquatic environments, and this is likely to be an advantage to individuals and to the species. Animals living in salt-water have different problems from those living in freshwaters. So it is more sensible to consider how animals cope with ionic and osmotic 'stress' by looking at how various unrelated animals cope in a particular habitat (e.g. freshwaters) rather than taking a 'taxonomic' view and dealing with animals group by group. A related problem is how animals are able to excrete the large amounts of nitrogenous products (e.g. ammonia) that are the end-products of amino acid metabolism. The best studied excretory organ is the human kidney and this remarkable organ is also the main site of osmotic and ionic regulation. Other renal organs have a quite different structure and function but, as is often true in physiology, we know a good deal more about the mammalian (and especially human) systems. However, there is currently much interest in non-mammalian systems, for example excretion in fishes and reptiles.

You will also learn something of how the kidney functions. Whatever the precise mechanism, the operation of the kidney is very complex, and in this and other fields our knowledge is incomplete.

The final Units (24 and 25) are concerned with the various ways in which animals acquire food and break it down physically and chemically into a form that can be absorbed. The type and quantities of chemicals that different animals require form the subject of nutrition. The mechanisms that animals use to gather and process food are very diverse, but we can more easily organize our knowledge by recognizing general categories that depend on the size and composition of the food taken in. For example, animals that feed on relatively small particles often display common features, even though they may be unrelated taxonomically. Keep in mind that such a system of classification, however good, is unnatural and sometimes arbitrary. Note too that the conclusions we draw are *generalizations*, not scientific laws; physiologists frequently meet animals that provide exceptions to the rules!

Much of our concern with the study of digestion centres around the activity of digestive enzymes, which break down foodstuffs very effectively. Because enzymes are specific, many different ones are necessary to process foodstuffs; animals that have a restricted diet have a narrower range of enzymes. Foodstuffs that are resistant to chemical and physical breakdown (especially the cellulose of plants) are a particular problem, and the solutions that animals have evolved are highly ingenious! Once the process of chemical breakdown is well advanced, the products of digestion can be absorbed; this topic is covered in Units 24 and 25.

Before you launch into what we hope will be an enjoyable and instructive block of Units, two points deserve emphasis. First, to organize this vast topic on a Unit basis, we have adopted a 'systems' approach, in which the function of one system (e.g. the circulatory system) is considered separately from another (e.g. the excretory system)—in reality all these systems are very much interrelated. While reading these Units, you should try to build up a picture of the functioning of the *whole animal*, which is more complex and interesting than merely the sum of all the individual components. Second, you will be presented with detailed information illustrating the key principles and generalizations that should act as focal points in your understanding. The Unit Objectives and the text summaries should help you to keep the principles in mind, but try not to become overwhelmed by detail; keep both the wood and the trees in view!

Planning your study time

Each of the Units in this block is equivalent to approximately one week's normal study time, except Unit 16 which is around one and a half Unit equivalents. In planning your study, look through the study guides to each of the Units.

Some of the Units (e.g. Unit 16) rely on knowledge of Foundation level biology, and you may have to revise the relevant material. Many of the Units build on knowledge of topics covered earlier in the Course (e.g. in Units 8–10), and you should therefore have the earlier Units to hand.

The television programmes, which run parallel to the Unit material, are referred to in the individual Unit study guides.

TABLE 1   A summary of vertebrate neurotransmitters and hormones grouped according to source

*(**Do NOT** attempt to memorize this Table—the list is for reference only.)*

Substance	Abbreviation	Chemical nature	Size	Specific source	Target	Action
1 NERVOUS SYSTEM						
acetylcholine	ACh	amine		CNS, PNS, ANS	local	neurotransmitter
dopamine	dopam.	amine		CNS, PNS, ANS	local	neurotransmitter
noradrenalin	nor adr.	amine		CNS, PNS, ANS adrenal medulla	local and diverse	neurotransmitters; contraction/relaxation of smooth muscle; alter tissue metabolism and stimulate glycogenolysis
adrenalin	adr.	amine	relative molecular mass 100–250	PNS, ANS adrenal medulla	local and diverse	
5-hydroxytryptamine (serotonin)	5HT	amine		CNS, mast cells, platelets	local	neurotransmitters; vasoconstriction/ muscle contraction
histamine	—	amine		CNS, mast cells	local	
melatonin	—	amine		pineal gland	hypothalamus + pituitary	promotes/inhibits gonadotropin release
γ-aminobutyric acid	GABA	amino acid	relative molecular mass 103	CNS	local	inhibitory neurotransmitter
glycine	Gly	amino acid	relative molecular mass 75	CNS	local	inhibitory neurotransmitter
2 HYPOTHALAMUS						
corticotropin-releasing factor	CRF	small peptide	?		ACTH cells in adeno-hypophysis	
thyroid-stimulating hormone releasing factor	TRF	small peptide	3 AAs		TSH cells in adeno-hypophysis (also PRL cells?)	
follicle-stimulating hormone and luteinizing hormone releasing factor	FSH/LH-RF	small peptide	10 AAs		ovaries and testes	
growth hormone release-inhibiting factor (somatostatin)	GIF	small peptide	14 AAs	(TRF and GIF also found in pancreas and intestine; GIF found in other neurons)	GH cells in adeno-hypophysis	action as implied by name (GIF and TRF also modify the activity of A and B cells in the pancreas)
growth hormone releasing factor	GRF	small peptide	10 AAs		GH cells in adeno-hypophysis	
prolactin release inhibiting factor	PIF	small peptide	?		PRL cells in adeno-hypophysis	
prolactin releasing factor	PRF	small peptide	?		PRL cells in adeno-hypophysis	
melanocyte-stimulating hormone release-inhibiting factor	MIF	small peptide	3 AAs		MSH cells in adeno-hypophysis	
melanocyte-stimulating hormone releasing factor	MRF	small peptide	5 AAs		MSH cells in adeno-hypophysis	

Substance	Abbreviation	Chemical nature	Size	Specific source	Target	Action
3 PITUITARY GLAND						
growth hormone (somatotropin)	GH	polypeptide	$\approx$ 190 AAs	adeno-hypophysis	all tissues	stimulates liver to form somatomedins, which alter tissue metabolism (liver, muscle, adipose tissue)
prolactin (luteotropic hormone)	PRL	polypeptide	$\approx$ 190 AAs	adeno-hypophysis	mammary gland	stimulates milk production
					corpus luteum	maintains secretion of oestrogen and progesterone
					liver	in some mammals, stimulates production of a pheromone that controls maternal behaviour
					fish gills	involved in maintenance of salt balance and osmoregulation
adrenocorticotropin (corticotropin)	ACTH	polypeptide	39 AAs	adeno-hypophysis (also found in in other regions of the brain)	adrenal cortex/ neurons	stimulates synthesis and release of glucocorticoids
melanocyte-stimulating hormone (intermedin)	MSH	small peptide	13–18 AAs	adeno-hypophysis	melanophores (cells containing black pigment)/ neurons	causes the dispersal of the pigment contained in melanophore cells, which darkens skin in lower vertebrates
β-lipotropin	β-LPH	polypeptide	$\approx$ 90 AAs	adeno-hypophysis	adipose tissues	mobilizes lipids
enkephalins	—	small peptides	5 AAs	adeno-hypophysis (also found in intestinal neurons)	central nervous system (CNS)	small peptides structurally related to β-LPH that display opiate-like activity, important in the CNS where they affect the firing rate of neurons; can also produce analgesia
endorphin	—	polypeptide	30 AAs		central nervous system (CNS)	
thyroid-stimulating hormone (thyrotropin, thyrotropic hormone)	TSH	2 subunit polypeptide	$\approx$ 209 AAs	adeno-hypophysis	thyroid gland	stimulates synthesis and secretion of thyroxin and triiodothyronine (see THYROID GLAND below)
follicle-stimulating hormone	FSH	2 subunit polypeptide	$\approx$ 250 AAs?	adeno-hypophysis	seminiferous tubules	stimulates production of sperm/maturation of sperm
					ovary (follicles)	stimulates follicle maturation
luteinizing hormone (interstitial cell-stimulating hormone)	LH	2 subunit polypeptide	$\approx$ 216 AAs	adeno-hypophysis	testis	stimulates synthesis and secretion of androgens (e.g. testosterone)
					ovary	controls final maturation of follicle, and ovulation; stimulates oestrogen and progesterone secretion and is involved in formation of the corpus luteum
oxytocin	—	small peptide	8 AAs	neuro-hypophysis (also found in other areas of brain)	uterus, mammary gland (affects neurons in brain)	causes contraction of smooth muscle; milk secretion
antidiuretic hormone (vasopressin)	ADH	small peptide	8 AAs		kidney (affects neurons in brain)	causes reabsorption of water, antidiuretic action
					adeno-hypophysis	stimulates release of hormones from pituitary

Table 1 (*cont.*)

Substance	Abbreviation	Chemical nature	Size	Specific source	Target	Action
4 PANCREAS						
insulin	—	polypeptide	51 AAs	B-cells	all cells	lowers blood glucose; alters protein and lipid metabolism
glucagon	—	polypeptide	29 AAs	A-cells	liver	increases blood glucose by increasing glycogenolysis
gastrin	—	polypeptide	17 AAs	D-cells (also found in stomach)	stomach	stimulates acid secretion
5 THYROID GLAND						
calcitonin	—	polypeptide	32 AAs		bones, kidney	inhibits calcium release from bones; decreases blood calcium levels
thyroxin	T_4	iodinated amino acid	relative molecular mass 650–800		most tissues	alter tissue metabolism and are involved in differentiation; increase metabolic rate in mammals, are involved in the control of metamorphosis in amphibians
triiodothyronine	T_3	iodinated amino acid			most tissues	
6 PARATHYROID GLAND						
parathyroid hormone (parathormone)	PTH	polypeptide	84 AAs		bones, kidney	elevates blood calcium and phosphorus levels by stimulating the mobilization of calcium from bones; inhibits calcium excretion from kidney
7 BLOOD						
erythropoietin		polypeptide	?		bone marrow	increases erythrocyte production
angiotensin I		small peptide	10 AAs	angiotensino-gen, a protein found in blood, is converted to angiotensin by renin, an enzyme released from the kidney	adrenal cortex	
angiotensin II		small peptide	8 AAs		adrenal cortex	stimulates aldosterone synthesis and secretion
8 SMALL INTESTINE						
enterogastrone		(this hormone has not yet been isolated chemically)			stomach	inhibits acid secretion in stomach
cholecystokinin	CCK	polypeptide	33 AAs		gall bladder	stimulates contraction of the gall bladder and release of bile salts
					pancreas	stimulates secretion of digestive enzymes (also exhibits cross-reactivity with gastrin)
secretin	—	polypeptide	27 AAs		pancreas	stimulates secretion of water and inorganic salts
9 ADRENAL CORTEX (For adrenal medulla, see nervous system.)						
glucocorticoids, e.g. cortisol (hydroxycortisone), corticosterone	—	steroid	relative molecular mass 300–370		most tissues	alter tissue metabolism (liver, muscle); stimulate conversion of proteins to amino acids, particularly important in fasting or hibernating animals; have anti-inflammatory action and appear to be involved in the termination of pregnancy
aldosterone	—	steroid			kidney	resorption of Na^+ and K^+ in kidney, sweat, and salivary glands, gut, amphibian skin and bladder, fish gills

Substance	Abbreviation	Chemical nature	Size	Specific source	Target	Action
10 TESTES						
androgens e.g. testosterone, 5α-dihydroxy-testosterone	—	steroid	relative molecular mass 300–370		most cells	development and maintenance of male characteristics and behaviour
11 OVARY						
oestrogens e.g. oestradiol-17β	—	steroid	relative molecular mass 300–370		most cells	development and maintenance of female characteristics and behaviour
12 CORPUS LUTEUM						
progesterone	—	steroid	relative molecular mass 314		uterus, mammary gland	maintenance of uterine endometrium and stimulation of mammary duct formation; acts in conjunction with oestrogen to maintain oestrus and menstrual cycles

AA = amino acid
PNS = peripheral nervous system
CNS = central nervous system
ANS = autonomic nervous system

Further reading

A list of further reading is given at the end of most of the Units. Here we mention text-books that cover wide areas of physiology, which you may like to consult. In particular, we recommend:

Schmidt-Nielson, K. (1979) *Animal Physiology, Adaptation and Environment*, Cambridge, 2nd edn. Hardback £32.50, paperback £9.50.

Hill, R. W. (1976) *Comparative Physiology of Animals: An Environmental Approach*, Harper & Row. £17.50.

Eckert, R. and Randall, D. (1978) *Animal Physiology*, Freeman. £14.10.

unit 16

Communication: Nerves and Hormones

Contents

TABLE A   Scientific terms and principles used in Unit 16

Assumed knowledge†	Introduced in an earlier Unit	Unit	Introduced or developed in this Unit	Page
acetylcholine (ACh)	allostery	7	ablation and reimplantation experiments	38
adrenalin	antibody	5	acetylcholine (ACh)*	17
allostery	autoradiography	4	action (spike) potential*	11
anion	bilateral symmetry	1	adenylate cyclase*	53
adenosine 5′-triphosphate (ATP)	cell membrane	4	adrenal gland*	38
axons	chromatin	12 & 13	adrenalin	38
capillaries	concentration gradient	9 & 10	adrenocorticotropic hormone (ACTH)*	46
carbohydrates	conductance G	9 & 10		
catalysis	cnidarians	1	afferent (sensory) nerve	26
cation	cnidoblast	1	agonist*	52
chemical equilibrium	cristae	4	amines	47
DNA	decarboxylation	8	androgens	31
electrical conduction[2]	diffusion	9 & 10	antagonist*	52
endocrine gland	electrochemical equilibrium	9 & 10	anterior lobe of pituitary* (adenohypophysis)	42
endoplasmic reticulum	embryo	11		
enzymes	endoplasmic reticulum	4	antibodies	36
fatty acid	enzymes	6	antidiuretic hormone (ADH)	32
feedback regulation[5]	exocytosis	4	autonomic nervous system (ANS)*	27
Golgi body (apparatus)	feedback regulation	6 & 7	APUD cells	47
hormones	Golgi body	4	axon hillock	7
hypothalamus	histochemistry	4	axons*	6
lipids	histones	5	axoplasm	10
macromolecules	hormones	12 & 13	axoplasm resistance R_a	14
messenger RNA	immune system	5	bioassay*	35
mitochondria	insertion and aggregation of proteins	9 & 10	blood–brain barrier	22
muscles	ion channels	9 & 10	calmodulins*	54
neurons[1]	ionophores	9 & 10	cell membranes	9
nucleotides	ion pumps	9 & 10	central nervous system (CNS)*	27
nucleus	K_d, half maximum transport rate	9 & 10	coding and integration*	24
parasympathetic nervous system[4]			concentration gradient	9
peptides	lipids	5	conductive region (of neurons)	7
pH	macromolecules	5	corticosteroids	31
pheromones	membrane potential E_m	9 & 10	cross-reactivity of hormones	35
polysomes	metabolites	8	cyclic nucleotides (e.g. cAMP)*	52
proteins	mitochondria	4	dendrites*	6
radioactive labelling	myelin	5	dendritic field	8
ribosomes	Nernst equation	9 & 10	dendritic spines	7
stretch receptor	osmotic pressure	9 & 10	depolarization*	10
sympathetic nervous system[3, 4]	oviduct	12 & 13	effector region (of neurons)	8
synapses	peptides	5	efferent (motor) nerve	26
transcription	permeability	5	electrical conduction	10
transmitter substances	phagocytosis	4	electrical resistance	10
vagus nerve[4]	phosphorylation	8	electrochemical equilibrium*	9
	precursor	5	endocrine glands*	37
	protein kinases	8	exocytosis	19

Introduced in an earlier Unit	Unit	Introduced or developed in this Unit	Page	Introduced or developed in this Unit	Page
resistance R	9 & 10	exocrine glands	37	posterior lobe of pituitary* (neurohypophysis)	42
ribosomes	4	fatty acids	31	precursor (parent) molecule* (e.g. prohormone)	33
steroid hormones	12 & 13	feedback regulation*	45		
transduction	5	first messenger*	50	progestins	31
transport molecules	9 & 10	'free' hormone*	34	prolactin (PRL)	36
voltage clamping	9 & 10	ganglia (*sing.* ganglion)	27	protein kinases*	53
		glial (e.g. Schwann) cell*	15, 21	quantal release theory*	19
		gonadotropins*	43	radioactive labelling*	36
		growth hormone (GH)	32	radioimmunoassay*	35
		half-life of hormones*	34	receptive region (of neurons)	7
		hormone-binding proteins in plasma*	34	receptors*	6, 50
		hormone polymorphism	32	reflex arc	26
		hormones*	31	regulation of receptors (sensitivity)*	51
		hyperpolarization*	10	refractory period*	11
		hypothalamus*	40	releasing and release-inhibiting factors (neurohormones)*	44
		immune system	36		
		ion channels*	12	saltatory conduction*	15
		ionic conductance G	12	second messenger*	50
		K_d, measure of affinity of binding*	50	somatostatin (growth hormone release-inhibiting factor)*	47
		local potentials*	11	steroid hormones*	31
		messenger molecules*	5	sympathetic and parasympathetic nervous systems*	27
		membrane potential E_m*	7		
		membrane resistance R_m	14	synapses*	5, 8
		myelin*	15	synaptic cleft	8, 19
		neurohaemal organs	41	synaptic vesicles*	8
		neurohormones*	41	synergism*	56
		neurons*	6	target (effector) cell*	5
		neurosecretion*	41	tetraethylammonium ions (TEA)	13
		neurotransmitters*	6	tetrodotoxin (TTX)	13
		node of Ranvier	15	threshold value*	11
		noradrenalin	38	thyroid gland*	34
		oestrogens	31	thyroid-stimulating hormone releasing factor (TRF)	44
		oxytocin	32		
		pancreas*	31	thyroid-stimulating hormone (TSH)*	42
		paraneurons or paracrine cells*	48	thyroxin	32
		peptide hormones*	32	transcription	53
		peripheral nervous system (PNS)*	27	transduction*	6, 23
		permeability	9	tropic action*	42
		phosphodiesterases	55	vesicular hypothesis of transmitter release*	19
		pituitary gland (hypophysis)*	41	v_{max} (B_{max}), density of binding sites*	50
		postsynaptic potentials (PSPs)*	20		

*These terms must be thoroughly understood—see Objective 1.

†Most of these terms are explained in the Science Foundation Course. Those that are of particular importance to the understanding of this text appear with a superscript number and have a full reference at the end of the Unit.

Study guide for Unit 16

This is the first of three Units (16–18) concerned with the regulation and control of physiological processes. In Unit 16, which is the equivalent of 1.5 Units, the basic principles underlying the major control systems are described and illustrated with specific examples. Units 17 and 18 take a more detailed look at the control of blood glucose levels and reproduction in mammals.

The main theme of this Unit is regulation and control. However, you will also find many examples of structure–function relationships where particular regulatory systems are considered in detail. The Unit is concerned primarily with nervous and hormonal control and builds on your knowledge of S101*, Unit 22, *Physiological Regulation* (S100†, Unit 18, *Cells and Organisms*). In this Unit, we outline the principles on which communication between cells is based.

The general principles exemplified in this Unit also apply to invertebrate nervous and endocrine systems, although they are derived here from studies on vertebrates. The TV programme associated with this Unit, *Insect Hormones: The Control of Moulting*, looks at the role of hormones in insect moulting.

Initially, the nervous and hormone systems are treated as though they are quite distinct (Sections 2–6). However, there are many reasons, as you will see later, why these two systems should not be treated in this way. The interdependence and similarities of the two systems are dealt with in Sections 7 and 8.

Sections 3 and 4 rely heavily on Units 9 and 10, Sections 2.2 and 4, so first quickly re-read these, or have the Unit to hand for reference purposes. The AC band *Transmission Along Axons and Across Synapses*, should help you through Sections 3 and 4; you could work through it in parts, as indicated in the text, or wait until the end of the Unit and use it for revision.

Because Unit 16 lays the foundation for Units 17 and 18, you may find the list of new terms in Table A formidable. Concentrate on those marked with an asterisk (see Objective 1); you will find many of the terms become more familiar as you work through later Units.

A list of all the chemical messengers that you will encounter in this block is in the Introduction to Units 16–25. Do not attempt to learn the details but use the list to remind yourself of details when you come across chemical messengers in later Units and when revising. The details are set out so that it is easy to see where a particular chemical message is released from, what the target is, and how the message affects the target.

You are advised to work steadily through the Unit and make sure you fully understand the basic principles. With this knowledge you will then be well equipped to deal with Units 17 and 18 and later physiological concepts. A possible work plan might be: Sections 1 and 2—half an hour; Sections 3 and 4—one and a half hours each; Section 5—one hour; Sections 6, 7 and 8—one and a half hours each. If short of time, you could skip Sections 5.1 and 5.2.

1 The blueprint for a communication system

An extremely important feature of all living organisms is their ability to respond to changes in both their external environment (e.g. fluctuations in temperature, pressure, heat, light, etc.) and internal environment (e.g. changes in blood glucose after a meal). In a multicellular animal, the response must occur in a coordinated manner and may involve only certain cells or tissues. It is therefore necessary to have an intercellular communication network that can translate the 'environmental' information received by special sensory cells into a form that can be handled and directly interpreted by other cells within the organism.

For centuries the nervous system has been recognized as being involved in regulatory processes. The Roman physiologist Galen thought that the brain converted

---

* The Open University (1979) S101 *Science: A Foundation Course*, The Open University Press.

† The Open University (1971) S100 *Science Foundation Course*, The Open University Press.

'vital spirits' in the blood into 'animal spirits' which were then carried along the nerves to all parts of the body, inducing movement in muscles and 'sensations' in body organs. The sixteenth century anatomist Vesalius, way ahead of his time, raised the question as to whether these 'animal spirits' were some substance that was transported along the nerve or a quality (like sound or light) that was transmitted along the solid matter of the nerve. A nineteenth century physiologist would have confidently asserted that the latter was the correct answer, because by this time it was known that nerves communicated rapidly via electrical 'impulses'. Through the early part of this century, evidence was beginning to accumulate that these electrical impulses stimulated the release of chemical substances (neurotransmitters) from the *synapses* (nerve endings). These neurotransmitters are then responsible for altering the activity of other nerves or organs. At the same time, another communication and control system was discovered which involved the release into the bloodstream of chemical substances (hormones) from specialized glandular tissues (endocrine glands). The bloodstream acts as a transport system for these hormones, so that an endocrine gland in one part of the organism can influence the activity of cells in another part. Animals thus possess two control systems, a nervous system and a endocrine system (Figure 1). In general, control

synapses*

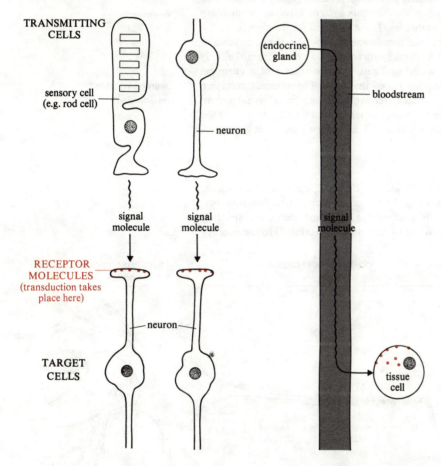

FIGURE 1 Various situations in which chemical communication is employed. All three systems use signalling substances (messenger molecules) but act over varying distances.

by nerves is extremely rapid, short in duration, and the response is precisely located; hormonal effects are slower in developing, longer in duration, and often involve widely scattered cells. Usually, both systems involve ultimately the secretion of chemical substances (*messenger molecules*). However, in conventional nerves, neurotransmitters act over very short distances (15–100 nm), whereas hormones act over much longer distances (from millimetres to metres). The nervous system is ideally suited for coordinating activities that involve rapid or continuous adjustments (e.g. breathing or running), whereas the endocrine system is best suited for relatively long-term regulation (e.g. the control of reproductive cycles—see Unit 18; or the maintenance of blood glucose levels—see Unit 17).

messenger molecules*

Although, superficially, these coordination systems appear to be very different, they share one common feature: the final transfer of information to the *target cell* involves specific chemical compounds. The basic intercellular 'language' is thus chemical. In essence, this language can be fairly simple because the chemical messenger molecule need only 'switch on' or 'switch off' a particular cellular process to produce profound effects. In recent years biologists have concentrated

target (effector) cell*

on trying to elucidate the mechanisms by which target cells are switched on and off. These studies have revealed a remarkable uniformity in the mode of action of both neurotransmitters and hormones. Both sets of chemical messengers achieve their effects by highly specific interactions with macromolecules (*receptors*) on or within the target cells. The binding of the messenger molecule to the receptor induces a conformational change that triggers a range of cellular events.

receptors*

The idea of two distinct but complementary control systems is a useful way of understanding the basic features of these two systems, and we shall therefore first discuss each system separately. However, modern physiologists would not want to draw a sharp distinction between the two systems, for various reasons; we shall return to this point in Sections 7 and 8.

2 Nervous systems: cells and processes

You should recall that the nervous system is composed of cells called *neurons* which communicate, one to another, via long processes called *axons* and *dendrites*[1]. The basis of signalling in the nervous system involves momentary changes in the permeability of membranes. These lead, in turn, to changes in the distribution of ions across the neuronal membrane that result in the generation of electrical impulses. These impulses normally travel one way only (from the body of the neuron to the tip of the axon) and result in the release of a chemical 'signal'—a *neurotransmitter*—from the tip of the axon. The adjacent receiving membrane (dendrite or neuron) *transduces* (converts) this chemical signal into an electrical impulse, and the whole process is repeated. Essentially, that is it: the electrical impulse is generated, carried, received, and a new impulse is generated.

neurons*
axons*
dendrites*

neurotransmitters*
transduction*

2.1 Neurons as simple cells

The internal structure of a neuron, as seen in Figure 2, consists of a nucleus, extensive endoplasmic reticulum, Golgi body and mitochondria. Neurons perform the basic functions of all cells, synthesizing macromolecules, transporting some and breaking down others. We shall return later (Section 7) to the concept

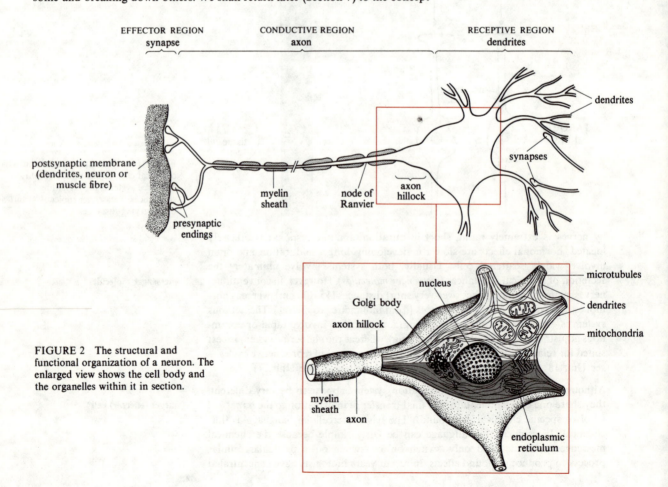

FIGURE 2 The structural and functional organization of a neuron. The enlarged view shows the cell body and the organelles within it in section.

6

of a neuron as a secretory cell. What then characterizes a neuron? Is it that they respond to environmental stimuli? You have already met several examples of non-neuronal cells with this property: for example, protistans will phagocytose particles they happen upon; the stimulation of many types of epithelial cells results in a discharge of mucus; touching the tentacles of a cnidarian triggers the release of cnidoblast threads that will trap prey (Unit 1, Section 4.1).

A neuron has a difference in *electrical potential* across the cell membrane (E_m) because of the uneven distribution of ions inside and outside the cell. Once again, this is not an exclusive characteristic of neurons: nearly all cells from every species ever looked at demonstrate this property (Units 9 and 10, Section 2.2). Ions are maintained at unequal concentrations across the membrane: penetrate any cell membrane with a microelectrode connected to a voltmeter and a small E_m can be measured (usually inside negative) because of the uneven distribution of, primarily, sodium and potassium ions (Na^+ and K^+).

membrane potential E_m *

Neurons do have very organized outgrowths called dendrites and axons, and these link up with the outgrowths of adjacent neurons to form a network. Once these processes are in close contact, the possibility exists for extensive cell-to-cell communication over long distances. The three-dimensional organization of neuronal processes is extremely varied and neurons can be classed according to the arrangement of these processes.

2.2 Nerve cells and their processes

The neuron is structurally bipolar: it has a dendritic zone and, at the opposite pole, an axon leaves the neuron body at the *axon hillock*. Figure 2 shows the structural and the functional polarity of a neuron. The dendrites form part of the *receptive zone* of the neuron and the axon *conducts* signals in the form of electrical impulses away from the cell body towards the synapse and the adjacent dendrites of other neurons.

axon hillock

receptive region (of neurons)
conductive region (of neurons)

Polarity is difficult to see in many neurons. Occasionally, the axon may divide, sending projections in different directions. Functionally, axons are concerned with the transmission of nerve impulses from the cell body to the synaptic region. As the derivation suggests (from the Greek for 'a tree'), dendrites are generally 'tree-like', although their shape varies in different neurons (Figure 3). The branching of dendrites may be simple, as in Figure 3a, or very complex, as on the neuron in Figure 3d. On many of the branches there are short projections or 'spines' which can be seen more clearly with the electron microscope.

dendritic spines

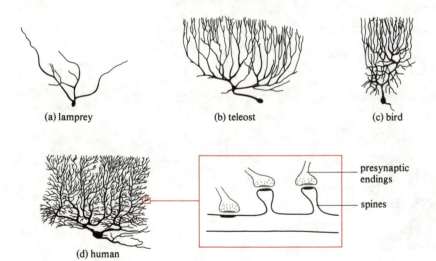

(a) lamprey   (b) teleost   (c) bird

(d) human

presynaptic endings

spines

FIGURE 3   Shapes of dendritic 'trees' on a type of neuron (Purkinje cell) from the cerebellum of different vertebrates. One dendrite is shown enlarged, and dendritic spines can be seen with incoming presynaptic endings.

☐ What could be the function of this extensive branching, and of the spines?

■ Dendrites receive inputs from synapses, and the large surface area provided by branching and spines increases the membrane area available for synaptic inputs.

Figure 4 shows the multiplicity of synaptic inputs that can be achieved on dendrites (certain large neurons can apparently receive up to 10^8 synaptic inputs).

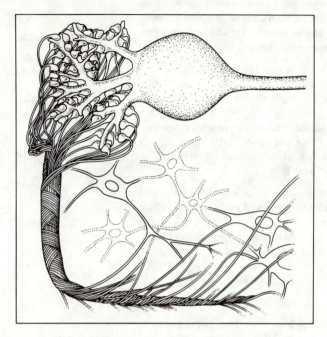

FIGURE 4 An artist's impression of the density of inputs to a single neuron.

The main *effector zone* of the neuron is situated at the end of the axon furthest from the cell body. The axon may either branch to form a number of synaptic swellings or develop such swellings along its length (Figure 5). A *synapse* consists essentially of the presynaptic ending of the axon, the postsynaptic region (which may be a dendrite on another neuron, or an adjacent nerve terminal, or a neuron cell body, or a muscle fibre, or an endocrine gland cell) and the small space between, the *synaptic cleft*. The inset in Figure 5 shows that the presynaptic swelling contains a number of spherical *vesicles* and that both the presynaptic and postsynaptic membranes are somewhat thickened. The significance of these various structures will become clear in Section 4.

effector region (of neurons)

synapses*

synaptic cleft
synaptic vesicles*

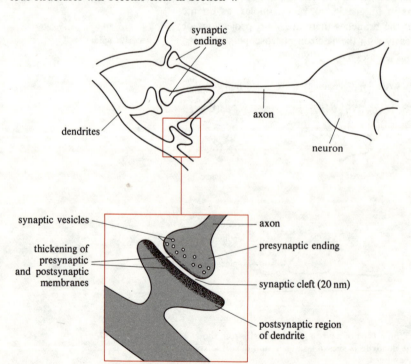

FIGURE 5 The major components of a synapse. An enlarged view of one synapse is shown in section.

The neuron can therefore be thought of as having receptive, conductive and effector regions (Figure 2). The relative extent of these regions varies, depending upon the type of neuron, and may imply something about the function of that particular neuron. The *dendritic field* (the extent of branching) can be enormous (as in Figure 3d): axons can run for several metres (e.g. the nerves in a giraffe's leg)

dendritic field

and eventually split into a network of terminals on numerous other neuron bodies or their dendrites. The final result is somewhat similar in arrangement to a huge, interconnected, telephone network (though there the analogy ends!). Before considering the implications of this synaptic organization within the nervous system, we shall look at the way in which signals are generated in the neuron and conducted along axons.

Objectives and SAQ for Section 2

Now that you have completed this Section, you should be able to:

★ illustrate with a simple diagram the structural and functional zones of a neuron.

★ define or recognize definitions of the following terms: polarity, dendritic spines, neuronal processes, synapses.

To test your understanding of this Section, try the following SAQ.

SAQ 1 (*Objective 2*) Figure 6 shows three neurons communicating with one another. Study the Figure carefully and select the *three* correct descriptions from (i)–(vii).

(i)   Signals pass from neuron N to neuron T via neuron X.

(ii)   U is a dendrite of neuron T.

(iii) X communicates with N via a synapse on a dendrite.

(iv)   Neurons T, X and N are distinctly polar.

(v)   T communicates with X by a synapse on a dendrite.

(vi)   Dendritic spines will be present on Y.

(vii) Signals pass from neuron T to neuron N, via neuron X.

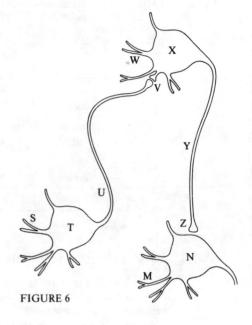

FIGURE 6

3 Nervous systems: signalling along axons

Communication in nerves is brought about by momentary changes in the permeability of membranes to various ions. Before you read this Section, make sure you understand what is meant by permeability, the equilibrium potential of an ion and an ion pump (Units 9 and 10, Sections 1.3, 2.2 and 4). The changes in the permeability of the membrane to ions that underlie the transmission of information along an axon are fairly complex, and the first part of the AC band, Transmission Along Axons and Across Synapses, is designed to help you through Section 3. The key points to remember are that axons possess voltage-dependent ion gates for Na^+ and K^+; these give rise to an explosive change in the E_m, called an 'action potential'. The action potential travels along the axon to the synapse.

From Units 9 and 10, you know that ions can cross *cell membranes*. In the example given in Units 9 and 10, Section 2.2.1, the membrane was assumed to be *permeable* to K^+ alone. If unequal concentrations of K^+ are present on either side of the membrane then K^+ will move down a *concentration gradient*. However, this movement is eventually opposed by an oppositely directed electrical force. A balance point is therefore reached (*electrochemical equilibrium*) which can be described using the Nernst equation.

cell membranes

permeability
concentration gradient

electrochemical equilibrium*

Living cells exhibit a membrane potential E_m which is often negative inside the cell with respect to the outside because of the presence of large anions which cannot cross the membrane. The value of this E_m will vary from one cell to another.

☐ Table 1 shows values for the membrane potential E_m across particular cells from three species—why are they different?

■ Part of the answer is given in the Table. The value of E_m depends upon the concentration of ions on either side of the membrane. This concentration varies between the cell types, but also depends upon the permeability of the membrane to particular ions.

For the squid axon in Table 1, we would predict from the concentrations of Na^+ and K^+, using the Nernst equation (Units 9 and 10, Section 2.2.1) that the E_m

would be $-75\,\text{mV}$ (inside negative) if the membrane was permeable only to K^+, and $+55\,\text{mV}$ (inside positive) if permeable only to Na^+. In fact, the E_m is $-60\,\text{mV}$ because the squid axon is relatively more permeable to K^+ than to Na^+. Note that E_m is only *slightly* less negative than the value for E_{K^+}. Any changes in the permeability to either ion will immediately change the E_m, and it is the degree of permeability to Na^+ and K^+ that is the key to the generation and propagation of signals in the nervous system.

TABLE 1   Ion concentrations ($\text{mmol}\,\text{l}^{-1}$) inside and outside various cells, and the membrane potentials E_m recorded when these cells are penetrated by a microelectrode

		Squid giant axon	Mammalian motor neuron	Frog muscle
K^+	inside	400	150	140
	outside	10	2.5	2.5
Na^+	inside	50	15	9
	outside	440	145	120
Cl^-	inside	40–100	9	3
	outside	560	101	120
	E_m/mV	-60	-70	-90

3.1   Local and action potentials

To start with, it is worth thinking of an axon in terms of an electric cable. The axon contains *axoplasm*, a solution of salts and proteins within a lipid–protein membrane. An electric cable consists of a metal wire contained within a plastic covering. If a wire is connected to the two poles of a battery a current will flow along it—the current being carried by the electrons in the wire[2]. If a thick piece of wire with a light bulb in the circuit is used to join the two poles of the battery, the bulb will give light of an intensity proportional to the current flowing in the wire. If a wire of lesser diameter is used, the bulb will give a light of lower intensity. Changing the diameter of the wire from thick to thin increases the resistance of the wire to the flow of current, and so less current flows across the filament in the bulb. If we used a very long piece of the thin wire and repeated the experiment we should see that length produces a further increase in the resistance to current.

axoplasm

So, *electrical resistance* is a function of the length and diameter of a piece of wire: resistance increases with length and decreases with larger diameter. The resistance of a wire also depends upon what metal that wire is made from. One last point to bear in mind. Wires are insulated to prevent short circuits. If an uninsulated wire comes in contact with another electrical conductor (e.g. water), then current may flow out of the wire at the point of contact.

electrical resistance

Let us see how an axon compares with an electric cable. The ions of the axoplasm are charged and therefore their movement generates a current, just like the movement of electrons in the wire. Axoplasm, however, is a poor conductor of electricity compared with a metal wire because the number of charge carriers is smaller and their mobility is less. *Conduction* of electrical currents in an axon would be further hampered because the membrane around the axoplasm is not a perfect insulator and therefore ions leak in and out of the membrane (Units 9 and 10, Section 2.1).

electrical conduction

Figure 7 illustrates some of the electrical properties of the axon. In this experiment an axon (in practice usually the axon of a large invertebrate, e.g. a squid giant axon) is impaled by three microelectrodes (Units 9 and 10, Section 3). Microelectrode P is used to pass a current (in a sense, to inject ions) across the membrane. The direction in which the current flows (which depends upon whether anions or cations are injected) will either displace the E_m ($-70\,\text{mV}$) towards zero (*depolarize* the membrane) or make the E_m more negative (*hyperpolarize* the membrane). The other two microelectrodes in Figure 7 (V_1 and V_2) record the E_m at different distances along the axon from P.

depolarization*
hyperpolarization*

In the first instance, a hyperpolarizing current A is passed through P and then some 20 ms later, a depolarizing current B is passed through P. Look at the E_m

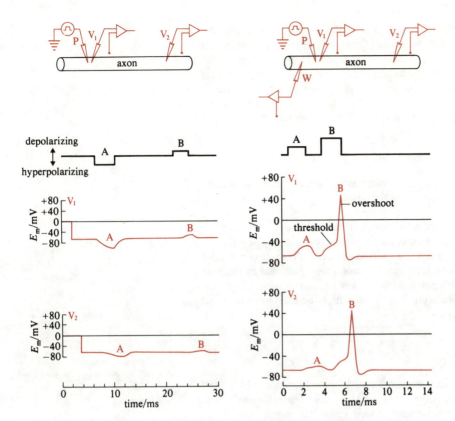

FIGURE 7   Measuring local potentials in a large axon. Δ = amplifier.

FIGURE 8   Recording an action potential in a large axon.

values recorded at V_1. The membrane is hyperpolarized and then depolarized as current flows from P to V_1. Now look at the record at V_2, a short distance away. The E_m at V_2 changes as current is passed at P but the changes in E_m are less, so the curve is much flatter.

The poor recording at V_2 is caused by the poor electrical properties of the axon. The small diameter of the axon and the leakage of current through the membrane means that current has a limited ability to flow along the axon; that is, the current attenuates (reduces in strength) over distance. This would not seem to be a propitious start for a communication system, for even if the signal managed to reach the target, it would bear no resemblance to what was originally transmitted. The experiment in Figure 7 shows that changes in the E_m decay over distance (and with time) along the axon because of its poor electrical properties—these changes are called *local potentials*. Obviously, if the axon of a cell was very short (the distance between P and V_1 in Figure 7), representative signals would be transmitted. Such local potentials may be very important in small neurons with short axons.

local potentials*

Now look at Figure 8. This time the membrane is progressively depolarized (A, then B) at P and the E_m is again recorded at V_1 and V_2. The first small depolarization A decays between V_1 and V_2, as in Figure 7, but the second *larger* depolarization B produces a massive change in the E_m, labelled an overshoot. This overshoot is faithfully reproduced at V_2 a few milliseconds later. This change is termed an *action* (or *spike*) *potential* and lasts for about 1.5 ms before the E_m is restored to the resting value. If a depolarizing current even larger than B is passed through P another action potential can be generated, but it would be *exactly the same size* as the last one. Thus, once the E_m has been pushed beyond a *threshold value*, an action potential of fixed size results—no more and no less. Transmission along axons is thus often termed an 'all or none' process.

If we were to attempt to generate a second action potential just 0.1 ms after the first, there would be no response. This is because, there is a period of enforced 'silence', called a *refractory period*, at the end of an action potential.

This refractory period results in the action potential travelling *away* from the point of stimulation because the membrane adjacent to the site of the stimulus is in an electrically receptive state, whereas the membrane *at* the site of the stimulus is in an electrically inactive state (Figure 9).

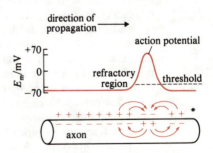

FIGURE 9   The refractory period of the action potential results in the next action potential (*shown by an asterisk*) being generated in an adjacent patch of membrane.

In the experiment shown in Figure 8 the change in the E_m is generated near the middle of the axon and so if we were to record at point W, we should see an action potential there as well. Normally, potentials are generated at the point where the axon leaves the neuron cell body (the axon hillock—the most excitable part of the axon) and will, therefore, pass down the axon towards the synapse. If the microelectrode is moved from V_2 to a position 1 metre down the axon, and a suitably large current is passed at P, we should still record an action potential at this distant point.

☐ How do these action potentials differ from local potentials?

■ Local potentials decay over distance whereas an action potential will be faithfully reproduced all the way along the axon.

The action potential is neither attenuated nor distorted because it is recreated in each successive patch of axonal membrane. Local potentials displace the E_m above the threshold value, and generate an action potential. This action potential then acts as a local potential and generates another action potential in the next patch of membrane. The action potential is the key to transmission. To throw light on what produces it, we shall return to the idea, first discussed in Units 9 and 10, of ion channels in membranes.

3.2 Ion gating and permeability

In Section 4 of Units 9 and 10 evidence was presented for the existence of both fixed (permanently open) and transient (open or closed) *ion channels* in cell membranes. Fixed channels provide for the movement of ions across cell membranes. Transient channels (ion gates) appear to be of two types: (a) those that are opened or closed by signal molecules such as hormones or neurotransmitters (simple ion gates), and (b) those that are operated by electrical fields (voltage-dependent ion gates). The latter channels are important in the generation of action potentials.

In the nervous system, as we shall see in the next Section, neurotransmitters open or close the simple type of ion gate. The resulting change in the permeability of the membrane (a) causes a new electrochemical equilibrium to be established leading to a change in the E_m and (b) may be sufficient to trigger voltage-dependent gates that in turn further displace the E_m, thus opening more gates, and so on. In a sense, the opening of the simple ion gates acts as a 'primer' to shift the E_m to a position (threshold) where the voltage-dependent gates are activated. Once these are activated, the result is an explosive change in the membrane's permeability to ions and, consequently, a rapid change in the E_m.

In Units 9 and 10, Section 4, you learned details of how three ion channels (ionophores) are created.

☐ Of the three ionophores, gramicidin, valinomycin and alamethicin, which was voltage-dependent?

■ Alamethicin. When a membrane containing this ionophore is depolarized there is an abrupt increase in current through the membrane.

Changing the E_m seems to reorient the alamethicin molecules, so that they take up a position across the membrane. In the nerve axon, a change in the E_m also leads to a sudden, massive depolarization. Can our knowledge of the way that alamethicin works tell us anything about the changes in the membrane that underlie an action potential? The major result of an action potential is that the inside of the axon becomes positively charged with respect to the outside. This must mean that there has been a redistribution of ions across the membrane, caused by either an *inflow* of cations or an *outflow* of anions, or both. In fact, the depolarizing effect of the action potential is caused by a rapid change in the membrane's permeability to Na^+. This can be shown by determining the *ionic conductance G* of the membrane to different ions during the action potential (see Figure 10). A change in the E_m, from $-70\,mV$ to $-60\,mV$, increases the permeability of the axonal membrane to Na^+ by some eight times. This in turn leads to the massive change in the E_m of $100\,mV$ (-60 to $+40\,mV$) as more of the voltage-dependent ion gates are opened.

ion channels*

ionic conductance G

12

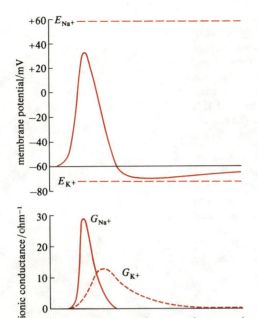

(a)

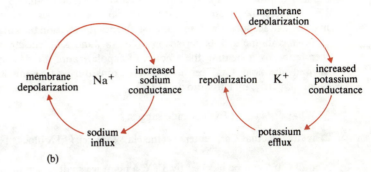

(b)

FIGURE 10 (a) Changes in membrane potential and ionic conductance during an action potential. (b) Changes in the conductance of sodium and potassium. The increase in sodium conductance is by positive feedback and in potassium conductance by negative feedback.

□ How does Na^+ enter the axon?

■ Na^+ simply moves passively down an electrochemical gradient into the axon.

The E_m at the peak of the action potential is $+40\,mV$, a value close to that for E_{Na^+} ($+55\,mV$).

□ What sort of feedback regulation is illustrated by this process of Na^+ voltage gating?

■ Na^+ gating is a positive feedback process—as the permeability increases, the E_m approaches zero, which increases the membrane's permeability to Na^+, and this further displaces the E_m.

Na^+ is therefore responsible for the depolarizing phase of the action potential. The end-point (or peak) is reached when the Na^+ gates close and K^+ gates are open to their greatest extent.

□ If the membrane's permeability to K^+ was increased, in which direction would K^+ move?

■ K^+ would tend to move out of the axon down a chemical gradient.

This outflow of K^+ restores the E_m to the resting condition; that is, it repolarizes the membrane. Note that the value for E_m closely approaches E_{K^+} ($-75\,mV$).

These sequential changes in permeability to specific cations during the action potential have been substantiated by measurements of Na^+ and K^+ fluxes in response to externally imposed changes of voltage (voltage clamping—a technique described in Units 9 and 10, Section 3). In addition, voltage-dependent ion gates for Na^+ and K^+ can be blocked specifically by *tetrodotoxin* (*TTX*—a poison) and *tetraethylammonium ions* (*TEA*), respectively (Units 9 and 10, Table 4).

tetrodotoxin (TTX)
tetraethylammonium ions (TEA)

The action potential is localized but passes along the axon as a wave of depolarization moving at a speed of some hundred metres per second. Propagation occurs because of the amplification effect of voltage-dependent ion gating. It is not yet known whether the gates in axonal membranes work by an aggregation system (as illustrated by alamethicin in Units 9 and 10, Section 4.1), but some of the properties of axonal membranes would fit such a model. For example, the insertion and aggregation of molecules would require a rather fluid membrane, and analysis shows that electrically excitable membranes have an unusually large proportion of unsaturated lipids (Unit 5, Section 8).

It is important to realize that the action potential results from the purely passive redistribution of ions. The consequence is that some Na^+ is gained by the axon and some K^+ lost, but only some hundred sodium ions traverse each ion channel during a single impulse and the total loss of K^+ is about 5 pmol cm^{-2} per impulse. The action potential is thus very economic. After a period of continuous firing of action potentials, the Na^+-K^+ exchange pumps in the axon are stimulated in order to restore the concentration of external Na^+ and internal K^+. This pumping is an active transport process and quite distinct from the passive processes involved in generating the action potentials.

So far, we have considered the role of Na^+ and K^+ in the transmission of signals along the axon, but other ions may be involved. Ca^{2+} plays a key role in a number of processes, and it has long been known that a reduction in the level of Ca^{2+} lowers the threshold for the initiation of impulses in both nerves and muscle. Extracellular Ca^{2+} appears to influence the relation between the potential of the membrane and its permeability to Na^+ and K^+, tending to stabilize the membrane by increasing the amount of depolarization needed to push the E_m beyond threshold. Ca^{2+} certainly enters axons during transmission and can be blocked, partly, by tetrodotoxin (TTX).

□ What does this TTX blockade suggest?

■ It suggests that Ca^{2+} enters via the Na^+ channel. (TTX blocks Na^+ channels.)

Not all Ca^{2+} can be blocked by TTX, a point we shall return to in Section 4. In some situations, Ca^{2+} can contribute to action potentials, for example in the muscle fibres of the heart and in some invertebrate neurons.

3.3 Fibre size, insulation and the speed of transmission

In the last Section we saw how local potentials, generated by the action potential, depolarize the membrane ahead of them, and if this results in an E_m above a certain threshold, yet another potential is generated. If a local potential is below the threshold, however, it will gradually attenuate along the axon. This decrement will depend upon the *electrical resistance of the axonal membrane* (R_m) and the *resistance of the axoplasm* (R_a). The passive spread of current along the axon is favoured if R_m is high relative to R_a because less current leaks out through the membrane.

membrane resistance R_m
axoplasm resistance R_a

□ What properties would you expect an axon to have if potentials were spread over long distances?

■ A high R_m and a low R_a.

Conversely, the current spreads over a shorter distance and attenuates more sharply in fibres with a low R_m and a high R_a. In practice, the size of the fibre is an important variable that affects how far the current spreads. The larger the fibre (e.g. the squid axon, diameter 0.5 mm), the further the local potential will spread. Most fibres and processes in the vertebrate nervous system, however, are only a few micrometres in diameter.

□ What effect would fibres of small diameter have on the spread of local potentials in vertebrates?

■ Local potentials will drop off very sharply over distance. This in turn influences the speed with which the change in local potential affects the adjacent patch of membrane.

Conduction of impulses in the axon is therefore considerably influenced by the resistance of both membrane and the axoplasm. The velocity of conduction is important for the way that the nervous system is organized and can vary by a factor of more than 100 between certain nerve fibres. Nerves that conduct most rapidly (more than $100 \, m \, s^{-1}$) are involved in reflex responses, for example the flip of a lobster's tail (an escape response) or the rapid withdrawal of a worm into its burrow or indeed, the jet-propelled escape of a squid. These three invertebrates have giant (large-diameter) axons that signal to the muscles involved in such fast reflexes.

So, in an axon with a high R_m and a low R_a the action potential proceeds rapidly along the axon because local potentials spread further. But what about the low R_m–high R_a fibres found in most vertebrates? In Unit 5, Section 8.2.1, we looked at the structure of the membrane of the Schwann cell, a *glial cell*. These cells wrap themselves around axons to give concentric layers of tightly apposed membrane called *myelin* (Figure 11). These membranes have a high resistance and therefore act as electrical insulators (increasing the R_m of axons). At intervals of about 1 mm, the myelin is interrupted, exposing a patch of naked, axonal membrane called the *node of Ranvier* (after its discoverer). As ions cannot flow into or out of the

glial (e.g. Schwann) cell*

myelin*

node of Ranvier

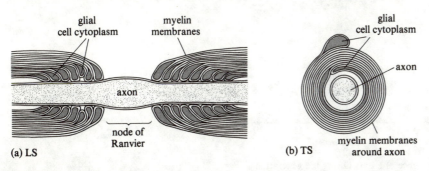

FIGURE 11   The myelination of an axon by a glial cell. At intervals (called nodes of Ranvier) the axon is bare of myelin.

myelinated (insulated) part of the nerve, during an impulse, current flows along the axon to the next node where an action potential is produced, and so on. Consequently, only a small proportion of the axonal membrane actually changes its permeability to ions, and the impulse travels rapidly in jumps from node to node—a mode that is called *saltatory conduction* (from the Latin for 'to jump')(see Figure 12).

saltatory conduction*

☐ What benefits would such insulation have for cell metabolism?

■ As only a small, nodal part of the axonal membrane is involved in transmission, changes in the distribution of ions are reduced and pump activity is kept to a minimum.

Myelination confers a distinct advantage on vertebrates and allows a 25-fold reduction in the diameter of fibres needed to attain a given velocity of conduction. Such small-diameter fibres are advantageous in lines of communication because more can be packed into a given space.

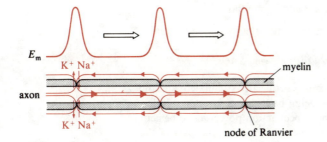

FIGURE 12   Saltatory conduction. Current flows between the nodes of Ranvier in a myelinated nerve; action potentials are generated at the nodes. Open arrows show the direction of movement.

Summary of Section 3

Look at Figure 13. The displacement of the E_m towards the threshold (1) can be achieved by local potentials. As the E_m reaches the threshold, voltage-dependent gates for Na$^+$ start to open (2–5) and, by positive feedback, the E_m value is driven explosively towards E_{Na^+} (5) (depolarization). As the action potential reaches its peak, Na$^+$ gates begin to close and K$^+$ gates begin to open (6): the E_m inverts and moves towards E_{K^+} (7–9). During the hyperpolarizing phase (6–10) the patch of

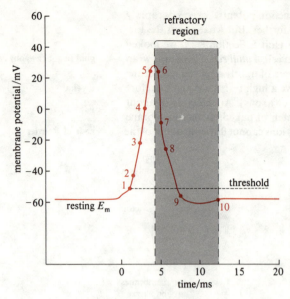

FIGURE 13 The sequence of events during the generation of an action potential: summary.

axonal membrane is refractory and cannot be depolarized. The change in the E_m stimulates the membrane just in front to depolarize and it in turn approaches the threshold. In this way, the action potential is propagated along the membrane. Subthreshold changes in the E_m (local potentials) simply decay with distance along the axonal membrane. The velocity of conduction is determined by the diameter of the axon and/or the presence of myelin. Myelinated nerves transmit impulses via saltatory conduction.

Before you start the next Section, you may find it helpful to work through the first part of the AC band associated with this Unit.

Objectives and SAQs for Section 3

Now that you have completed this Section, you should be able to:

★ describe the basis of the membrane potential of a neuron in terms of the distribution of ions and the permeability of the membrane.

★ distinguish between an action potential and a local potential in terms of the membrane's permeability to ions.

★ list the principal events in the course of an action potential.

★ describe the relationship between fibre size and the velocity of conduction.

To test your understanding of this Section, try the following SAQs.

SAQ 2 (*Objective 3*) An isolated squid axon is artificially stimulated, and the changes in the E_m are recorded while it is being bathed in sea-water solution containing a metabolic inhibitor. Would you expect action potentials to be generated: (i) for the whole period of the experiment, (ii) for the first part of the experiment or (iii) not at all? Explain your reasoning.

SAQ 3 (*Objective 4*) Examine Table 2.

TABLE 2 The velocity of conduction and the fibre size of selected axons from three species

	Diameter of fibre/μm	Velocity of conduction/m s^{-1}
species X	7	1.2
species Y	500	33
species Z	15	90

(a) Two of the three species are invertebrates. Which are they and why do you classify them in this way?

(b) How do you explain the fact that the velocity of conduction in species Z is greater than in species Y, but the fibre diameter is greater in species Y than in species Z?

(c) Why is the velocity of conduction in species Y greater than in species X?

SAQ 4 (*Objective 3*)   The E_m of a squid axon is normally $-60\,\text{mV}$. What would you expect to happen to the E_m immediately after each of (a)–(d) (i.e. would the axon membrane hyperpolarize or depolarize), and why?

(a) A sudden increase in permeability to Na^+

(b) A sudden increase in permeability to Cl^-

(c) Rupture of the membrane

(d) A rise in the concentration of K^+ outside the axon.

SAQ 5 (*Objective 3*)   What is the explanation for the fact that a local potential decays over distance along an axon whereas an action potential is faithfully reproduced all the way along an axon?

4   Nervous systems: transmission across synapses

In Section 3 we saw how impulses are transmitted from the neuronal cell body, along the axon, in such a way that the signal is faithfully reproduced. Between one neuron and another there exists a gap in this line of communication: effectively a break in the circuit. It is now accepted that the means of communication across the synapse is chemical—though, as we shall see, some synapses are so closely apposed that direct electrical transmission is possible.

At the turn of the century, there was heated debate between physiologists about the relationship between neurons at the synapse. The idea of chemical transmission was largely based upon anatomical and electrical observations and was not popular. Transmission across synapses was known to take place in a small fraction of a second, and to many it seemed more likely that direct electrical contact was involved. In 1921, Otto Loewi reversed this thinking dramatically! Loewi perfused* the heart of a frog and electrically stimulated the vagus nerve, which slows the heart-beat. He then transferred samples of the perfusate from this heart to a second one, and noted that the rate of this heart also slowed. 'Something' released by the vagus nerve produced a slowing of the heart, and Loewi went on to show that this 'something' in fact was *acetylcholine (ACh)*. Loewi admitted later that the idea for this experiment came to him in a dream and he wrote it down in the middle of the night. Unfortunately, the next morning he could not read his own writing, but when the dream returned, rushed immediately to his laboratory to perform the experiment.

acetylcholine (ACh)*

Evidence that acetylcholine was a chemical transmitter began to accumulate both from experiments on the autonomic nervous system[3] and on mammalian neuro-muscular junctions (the synapses between axons from the spinal cord and the skeletal musculature). Today (1980) a dozen or so compounds have been shown to act as neurotransmitters, and a whole host of others have been implicated but not yet proven. In Section 8 we shall look more closely at the nature of these transmitters and of signal molecules in general, but let us now consider how electrical impulses trigger the release of chemicals and how these subsequently trigger electrical activity in postsynaptic structures.

4.1   Presynaptic events

Many of the pioneering experiments were made on simple preparations, such as nerve–muscle synapses of the frog (see Figure 14). The transmitter involved is known to be acetylcholine and the large size of this synapse (compared with those in the brain, for example) makes it ideal for studying the presynaptic release of, and postsynaptic response to, this transmitter. If an impulse is passed down the

---

*Perfusion is the passing of a blood substitute (e.g. saline) through an isolated organ.

axon to the neuromuscular junction, it takes something like 0.5 ms after arrival at the synapse before a response can be recorded in the muscle fibre. This delay at the synapse is sensitive to temperature, for if the preparation is cooled to 2 °C, then the delay is increased to 7 ms. If acetylcholine was simply *diffusing* across the cleft from presynaptic to postsynaptic membranes, the delay need only be as short as 50 μs. Some metabolic event appears to be involved.

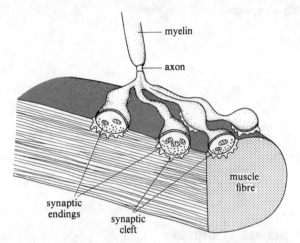

FIGURE 14 The neuro–muscular junction. The axon divides into a number of specialized endings that form synapses with a muscle fibre.

Ca^{2+} is involved in transmission (Section 3.2). If the concentration of Ca^{2+} in the fluid around the synapse is lowered, the release of acetylcholine is also reduced. The importance of Ca^{2+} for the secretion of neurotransmitters has been established at all chemical synapses, and its general role in stimulating secretion has been extended to other processes (e.g. the release of hormones; see Section 6).

☐ What is the effect of the depolarization of a membrane on its permeability to Ca^{2+}? (Think back to Section 3, on axonal transmission.)

■ Permeability to Ca^{2+} is increased with the depolarization of the membrane and Ca^{2+} enters the cell.

This link was elegantly demonstrated in 1971 by Bernard Katz and Ricardo Miledi working in London. They reasoned that if the presynaptic terminal was artificially depolarized (i.e. clamped) to the equilibrium potential for Ca^{2+} $(E_{Ca^{2+}})$ then no Ca^{2+} should enter as the action potential reached the synapse. And, in fact, no acetylcholine was released. So, the sequence of events appears to be: depolarization of the presynaptic terminal by the arrival of a nerve impulse, the entry of Ca^{2+} and the release of a chemical transmitter. But how is the transmitter released?

Two earlier findings had suggested that the release of the transmitter is not simply caused by an increase in the permeability of the presynaptic membrane to acetylcholine. First, even if no impulses were travelling down the axon to the synapse, small, spontaneous, 'miniature' potentials (of about 1 mV) could be recorded across the synapse at the postsynaptic membrane. Second, when the presynaptic ending was gradually depolarized, the miniature potentials recorded in the postsynaptic membrane increased in size proportionally (see Figure 15). It seems,

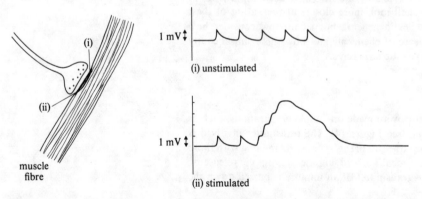

FIGURE 15 The quantal nature of transmitter release. In the unstimulated state (i) miniature potentials of fixed size are recorded in the muscle fibre. Stimulation (ii) results in a potential that comprises the sum of small, spontaneous potentials, which are now occurring with increased frequency.

therefore, that the release of the transmitter occurs in quanta (small packages of fixed size). In the quiescent state, the *spontaneous* release of such quanta gives rise to the *miniature potentials*. When the terminal is gradually depolarized, more and

18

more quanta are released. This *quantal release theory* of neurotransmission is now generally accepted, but the mechanism of release and exactly where the transmitter is packaged into quanta are both hotly disputed.

quantal release theory*

When examined by electron microscopy the presynaptic ending can be seen to contain a number of vesicle-like structures that would appear to be prime candidates for storing the transmitter substance. These vesicles (40–200 nm in diameter) can be separated from the synapse and do indeed contain a neurotransmitter; they can also be stained *in situ* with agents that are known to react with neurotransmitter substances. If a nerve is stimulated hard, so that much transmitter is released, the number of vesicles decreases. Conversely, when the nerve is recovering from such hard stimulation, the number of vesicles increases again. The release of transmitter is believed to be by *exocytosis* (Unit 4, Section 4.5). Vesicles fuse with the presynaptic membrane and empty their contents into the *synaptic cleft* (see Figure 16).

See *The S202 Picture Book* for presynaptic endings.

exocytosis
synaptic cleft

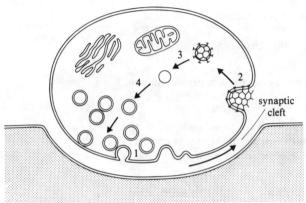

FIGURE 16   The vesicular hypothesis of transmitter release. 1   Vesicles fuse with the presynaptic membrane and release their contents. 2   Endocytosis of the membrane pinches off new vesicles. 3   New vesicles are formed. 4   Vesicles are ready to release a chemical transmitter.

Ca^{2+} is somehow involved in the fusion between vesicle and synaptic membrane, and depolarization of the membrane simply increases the frequency of this process. This *vesicular hypothesis*, as it has become known, is attractive, but by no means fully accepted. Microscopic evidence for exocytosis is scarce, and it is possible that the vesicles simply act as reservoirs for the transmitter. The release of a transmitter substance could be through specialized channels in the membrane such as those described in Section 3. When the action potential depolarizes the synaptic ending, transmitter is released from the cytoplasm into the cleft through voltage-sensitive channels—a process similar to those demonstrated for Na^+ and K^+ along axons.

vesicular hypothesis of transmitter release*

To summarize presynaptic events, look at Figure 17, which shows that an action potential arriving at the presynaptic terminal causes a release of transmitter, which reacts with receptors on the postsynaptic membrane to alter its permeability to ions (e.g. Na^+ and K^+). Impulses arriving at the presynaptic terminal promote the entry of Ca^{2+}, which triggers the secretion of the stored transmitter substance. The transmitter is secreted via a mechanism, as yet unknown, but apparently in a regulated, quantal form. The transmitter substance diffuses across the narrow synaptic cleft and interacts with the postsynaptic membrane, inducing a change in membrane potential.

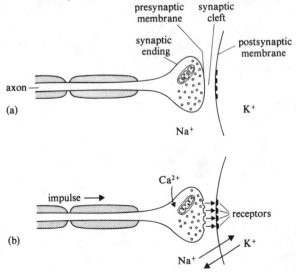

FIGURE 17   The sequence of events during synaptic transmission.

4.2   Postsynaptic potentials

The area of membrane (dendrites or cell body) directly under the presynaptic ending is generally believed to be incapable of generating action potentials. Instead, the postsynaptic membrane simply responds by changing its potential as a result of the presence of chemical transmitters in the synaptic cleft. This local change in potential is proportional to the quantity of neurotransmitter interacting with (i.e. bound to) the receptor. Such *postsynaptic potentials* (or *PSPs* as they are usually called) are generated by changes in the membrane's permeability to ions (as for the action potential), but the changes are due to the opening of simple (ligand-dependent) gates for ions rather than voltage-dependent gates.

postsynaptic potentials (PSPs)*

The neurotransmitter acetylcholine (ACh) is itself a positively charged molecule and would therefore depolarize the membrane if it entered the postsynaptic ending in any quantity. From the amount of current that flows through the postsynaptic membrane, though, it can be estimated that too little acetylcholine is released for it alone to account for the change in potential.

□ If the change in the potential of the postsynaptic membrane is not caused by ACh itself, what must be happening?

■ The change in potential must be a consequence of changes in permeability to ions *produced* by ACh.

The ion species involved in these postsynaptic potentials depends upon the particular synapse under investigation. In frog neuromuscular junctions, radioactive labelling of ions indicates that the membrane becomes more permeable to Na^+, K^+ and Ca^{2+} but not to Cl^-.

These changes in permeability to ions are not the same as those that underlie the action potential in which permeability to Na^+ increases first followed by an increase in permeability to K^+. In the postsynaptic membrane, the changes in permeability to ions are simultaneous and do not operate by voltage-gating.

□ Predict very roughly a value for the postsynaptic membrane potential if permeability to Na^+ and K^+ increases simultaneously?

■ If the increase in permeability is the same for both ions, then the E_m would take an intermediate value between E_{K^+} ($-74\,mV$) and E_{Na^+} ($+45\,mV$), that is about $-25\,mV$. (Recall the unstimulated E_m in rod cells from the TV programme, *The Rod Cell*.)

□ When measured, the PSP is actually close to zero. What does this suggest?

■ That permeability to Na^+ is increased more than permeability to K^+ because the E_m (about zero) is closer to E_{Na^+} ($+45\,mV$).

The differential permeability to Na^+ and K^+ varies at different synapses; for example, in the crustacean neuromuscular junction the value is $+20\,mV$. The important thing is that the resulting change in the E_m is sufficient (i.e. above threshold) to initiate an action potential in the membrane at the axon hillock (see Figure 2 on p. 6). Such postsynaptic potentials are termed *excitatory* postsynaptic potentials.

In some synapses there is an increase in permeability to Cl^-.

□ What would you predict would happen to the E_m if the permeability to Cl^- was increased?

■ The equilibrium potential for Cl^- (E_{Cl^-}) would be close to the resting potential, or just below it; that is, a *hyperpolarization* would result.

Because a hyperpolarization would make it harder for the next action potential to produce a large enough excitatory postsynaptic potential to reach the threshold, this hyperpolarization is termed an *inhibitory* postsynaptic potential (Figure 18).

The skeletal muscles of vertebrates (e.g. the frog) do not receive inhibitory inputs directly. In crayfish and lobsters, however, muscles have both excitatory and inhibitory synapses. The inhibitory synapses here involve increased permeability to Cl^-, which thus keeps the postsynaptic potential well below threshold. The action of acetylcholine on the mammalian heart is also inhibitory but this

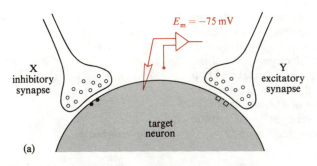 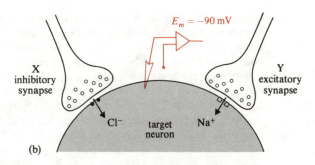

FIGURE 18 (a) A neuron receives both inhibitory (X) and excitatory (Y) synapses. (b) Input from X pushes the E_m away from the threshold and so it is harder for any subsequent input from Y to depolarize the neuron sufficiently.

is achieved solely by increasing the permeability to K^+. The E_{K^+} is close to the resting potential and well below the threshold. Acetylcholine can thus cause an excitatory or an inhibitory postsynaptic potential depending upon the particular synapse or, more correctly, depending upon the relationship between the receptor for acetylcholine on the membrane and the ion gates to which the receptor is linked. We shall look more closely at receptors in Section 8.

The duration of a postsynaptic potential is short because mechanisms exist to deactivate the chemical transmitter. For example, acetylcholine is rapidly degraded by an enzyme that is present in the synaptic cleft. Other transmitters are rapidly taken up by neighbouring glial cells and the presynaptic membrane, and are therefore deactivated by physical removal from the cleft. This uptake process is an active one and the sites of uptake have a very high affinity for the transmitters (Units 9 and 10).

A single neuron can receive many thousands of inputs, some inhibitory and some excitatory. The subsequent initiation of an impulse in the neuron depends upon the integration of conflicting inhibitory (hyperpolarizing) and excitatory (depolarizing) inputs, over both space and time. If the net result is a local potential of sufficient size to reach the threshold, then an action potential will be generated in the axon of the receiving neuron. If inhibitory inputs prevail, no action potential will be generated. This is the important function of synapses: they act as relay stations in the line of communication in the nervous system, summing up converging inputs and then either passing on the signals, or blocking them!

Direct *electrical* contacts have been described at a number of sites, although these are scarce compared with synapses that release chemical messages. The gap between membranes at chemical synapses is some 20 nm. At electrical synapses this gap is only 2 nm and current from the presynaptic ending flows across the gap and changes the potential in the postsynaptic membrane directly.

4.3 Glial cells

Most neurons are surrounded by cells that are collectively termed *glia*. In Section 3 we saw how one type of glial cell, the Schwann cell, is involved in myelinating (insulating) axons in some areas of the vertebrate nervous system. In the vertebrate brain, other types of glial cell can be recognized, for example, oligodendroglia (from the Greek for 'glia with few processes'), which perform a similar role to Schwann cells, spinning myelin membranes around axons.

glial (e.g. Schwann) cell*

Glial cells outnumber neurons in the brain by 10 to 1, but their role has in the past been dismissed somewhat summarily. Results now unfolding suggest there are many very complex interaction of neurons and glial cells.

If a glial cell is penetrated with a microelectrode, a large, stable E_m can be recorded (-70 to -90 mV), but no action potentials are observed. Synapses have never been seen on glial cells, which suggests there is little involvement of glia in neurotransmission. However, changes in the E_m of glial cells can be recorded when adjacent neurons are stimulated, and glial cells appear to control the concentrations of ions around neurons.

Recently, glial cells have been shown to function as regulators in other ways that involve taking up (or deactivating) the chemical transmitters released by synapses, and actually to provide axons with macromolecules synthesized within the glial cytoplasm. The function of this exchanged material is, as yet, unknown. Evidence for cooperation between glia and neurons is beginning to appear, and it is not unreasonable to assume that reverse traffic of macromolecules (from neuron to glia) may also occur.

Glial cells also play an important role in the general regulation of the environment around neurons. Blood vessels pervade the brain supplying it with glucose and oxygen. If a coloured dye is injected in the blood system, it will penetrate all body tissues *except* the brain and spinal cord. A careful examination of the capillaries in the brain shows that the dye passes through the cells that make up the capillary wall but will not pass through a layer of glial cells that lie alongside these small vessels. These glial cells (called astroglia), and to a certain extent the capillary cells, form a selectively permeable barrier between the blood and the cells of the brain, commonly called the *blood–brain barrier* (see Figure 19).

blood–brain barrier

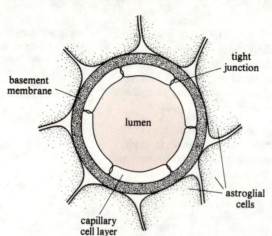

FIGURE 19 Elements of the blood–brain barrier. A brain capillary is formed by endothelial cells, which are connected by continuous belts of tight junctions, thus restricting diffusion between cells. The basement membrane and astroglial cell layer add to the barrier properties.

The blood–brain barrier is important in maintaining the environment around nerve cells: it protects them against toxins and metabolites in the blood and buffers the brain against wide fluctuations in blood constituents. At some places, notably around the hypothalamus, this barrier is not uniform; that is, it is somewhat leaky. The significance of this 'breach' in protection will become apparent in Section 7.

Summary of Section 4

Depolarization of the presynaptic membrane triggers the release of chemical transmitter into the synaptic cleft. The transmitter interacts with specific receptor sites on the postsynaptic membrane, which are linked to simple ion gates. Changes in the membrane's permeability to ions lead to alterations of the postsynaptic E_m that spread to the axon hillock. If the E_m is sufficiently displaced a new action potential will be triggered in the axon hillock. Transmitters are termed excitatory or inhibitory, depending on which ion gates they open. The change in the E_m represents the sum of all synaptic inputs and the initiation of a new action potential in the receiving neuron will depend upon the number and time of arrival of inputs to the cell.

Some electrical synapses are found; these communicate through the movement of ions across closely opposed presynaptic and postsynaptic membranes.

Glial cells form intimate functional relationships with neurons, deactivating transmitter substances by removing them from the synaptic cleft, insulating nerve axons, supplying macromolecules to neurons and forming part of the blood–brain barrier.

Before going on to the next Section, you may find it useful to work through the second part of the AC band associated with this Unit.

Objectives and SAQs for Section 4

Now that you have completed this Section, you should be able to:

★ describe the major events between the arrival of an action potential in an axonal ending and the generation of another in the axon hillock of a receiving neuron.

★ explain, in terms of changes in the permeability of membranes to ions, how some synapses are inhibitory and others excitatory.

★ list at least three examples of the relationship between neurons and glial cells.

To test your understanding of the Section, try the following SAQs.

SAQ 6 (*Objective 5*) What would be the result of bathing an isolated nerve–muscle preparation in tetrodotoxin (TTX) (which blocks the voltage-dependent Na^+ channel) (i) with TTX applied only around the nerve–muscle junction and (ii) with TTX around the axon? Would you expect to see postsynaptic potentials in the muscle after (i) or (ii) when the axon is artificially stimulated?

SAQ 7 (*Objective 5*) At synapse A a transmitter opens gates for Na^+ and K^+ whereas at synapse B the same transmitter opens just Na^+ gates. Which synapse will produce the larger change in the E_m, and in which direction will this be (hyperpolarization or depolarization)?

SAQ 8 (*Objective 6*) Select the *two* correct statements from (i)–(vi).

(i) The blood–brain barrier prevents all blood-borne substances except gases from entering the brain.

(ii) Large glial cells have numerous fine processes that receive synaptic inputs from neurons.

(iii) All glial cells, except Schwann cells, demonstrate a potential difference across the membrane (E_m).

(iv) Oligodendroglia and Schwann cells form insulating layers of myelin around nerve axons.

(v) Glial cells do not possess voltage-dependent ion channels.

(vi) Glial cells act *only* as regulators of the extracellular environment around neurons.

5 Nervous systems: coding and transmitting in circuits

The main concepts in this Section include the transduction of environmental stimuli by sensory cells, coding and integrating within the nervous system and functional divisions within the nervous system. If you are very short of time, concentrate on Section 5.3.

In Sections 3 and 4 we looked at the way that two neurons, or a neuron and muscle fibre, communicate with one another. In this Section we investigate where and how nervous impulses are generated (at sensory nerve endings) and how these are 'coded' and integrated with electrical activity in other pathways in the nervous system. In Section 5.3 we shall see how cells are organized into circuits, and circuits into systems. Functionally, the nervous system of an organism is not one but a collection of different systems interacting with each other to coordinate the activities of the organism.

5.1 Generating potentials in sensory cells

The output from a neuron is a function of various excitatory and inhibitory inputs arriving at the synaptic endings on dendrites or on the cell body. Some neurons are specialized to detect other types of input arising either from within the organism (e.g. the contracted or relaxed state of a leg muscle) or outside in its environment (e.g. a pin prick on the skin).

The organism's perception of its environment, and of its place within that environment, depends upon specialized, sensory nerve endings that function just like synapses. All sensory receptors are anatomically arranged to respond to mechanical, chemical or electrical (e.g. in some fish) stimuli or to light. A sensory process, then, involves translating the signal from one form (e.g. light acting on the retina of the eye) into another (e.g. the electrical events in the optic nerve to the brain), just as the synapses we have been considering convert chemical signals (e.g. ACh) into electrical activity. This conversion process is termed *transduction*.

transduction

Transduction always results in an action potential in an axon connected with the sensory cell, but the mechanism of transduction obviously varies according to the nature of the stimulus. Thus, a sensory cell specialized to perceive changes in the intensity of light, such as the retinal cell of the eye, generates potentials in a different way from the sensory nerve ending that responds to pressure in the skin on the back of your hand.

Figure 20 shows a sensory hair cell, which is one of thousands within a specialized receptive region of the skin of fishes. This cell responds to the force and direction of the water current along the body. If a microelectrode is positioned close to the sensory axon of the cell, electrical activity can be recorded. You can see that as the hairs on the cell are bent in one direction, the frequency of impulses travelling down the axon increases.

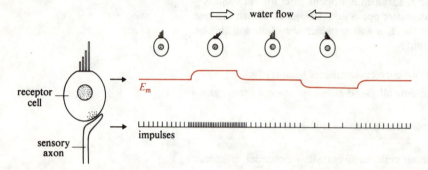

FIGURE 20   A sensory hair cell and its axon in the skin of fish. When the hair is bent there is a change in the E_m in the sensory cell. The lower line shows the frequency of nerve impulses that are recorded in the axon. (NB   This cell is unusual in that it communicates across a synapse with its axon.)

☐ Why is there a change in the frequency of action potentials rather than a change in their amplitude?

■ Remember that the action potential has a fixed amplitude. Once the threshold E_m is reached (the stimulus being in this example the deformation of the hairs of the cell), the action potential initiated is of fixed size, irrespective of any further increase in the intensity of the stimulus (Section 3.1).

Any change in the stimulus (e.g. in intensity, or 'direction' in this case) is thus coded for by a change in the frequency of action potentials in the axon. The coding of information in the nervous system is, and can only be, by changes in the *frequency* of nerve impulses, sometimes known as the rate of 'bursting'.

5.2   Coding and integrating information

Communication in the nervous system is achieved through the movement of ions but because of the 'all or none' properties of the action potential it would appear that communication must be limited. A receiving neuron will respond to the arrival of an action potential but will not be 'aware' of the intensity of the stimulus that generated that potential—nor indeed of the nature of the original stimulus. The amount of 'information' in one action potential is thus very limited. Information in the nervous system is *coded*, and because of the limits of the action potential, the only possible means of coding is according to the frequency of impulses. Where the inputs happen (i.e. where messages arise) is indicated by the way neurons connect with one another to form pathways and circuits. Finally, information is processed in the nervous system. To store information, and modify it in the light of all the other information available, there must exist sites in neural pathways where information is amplified or overruled and *integrated* with information arriving from other places. The sites that carry out integration are the synapses between neurons: a system of nerves without synapses could only perform stop–go functions.

coding and integration*

In the last Section you learned how information about water flow is coded in the skin of the fish. Now look again at Figure 20.

□ What coding characteristic does the receptor cell show when the water flows from left to right?

■ The receptor cell shows an increase in the frequency of impulses.

□ What codes for the reversed water flow?

■ A reduction in the frequency of firing, a consequence of the hyperpolarization of the membrane.

□ What is the electrical behaviour of the receptor cell when it is not being stimulated?

■ The axon of the receptor cell shows a constant rate of firing.

The receptor cell provides useful information about the stimulus because the frequency of firing varies above and below the base line for spontaneous firing. Spontaneous firing is a property of many sensory receptors and presumably results from the 'leakiness' of the membrane of the receptor cell. As ions leak in and out, the potential reaches threshold and an action potential is generated. A stimulus simply increases the leakiness, and the result is either a depolarization or a hyperpolarization.

□ What will dictate whether the membrane is depolarized or hyperpolarized?

■ The nature of the ion. If Na^+ is involved, then the potential will approach zero: if K^+ is involved or Cl^-, then the membrane will hyperpolarize.

Sensory receptors transduce environmental information and this is coded in a number of ways. The *onset* and the *end of a stimulus* is coded by changes in the rate of firing; the *strength* by the frequency (bursting); and the *duration* of the stimulus by the length of the period during which impulses are triggered (the duration of bursting).

The dendrites of a neuron can receive multiple inputs (recall Figure 4 on p. 8). If simultaneous excitatory and inhibitory inputs arrive, the response of the neuron will depend upon the size of the resulting change in potential, which in turn depends upon summation of the inputs. If the local potential reaches the threshold it will fire an action potential; if below the threshold, it will not. However, if a single excitatory input arrives a millisecond later than the first the potential will still be slightly displaced towards the threshold and the additional input may be sufficient to push the potential above the threshold. In this case the efficiency of the synaptic input depends on its timing; but it also depends upon where it arrives!

In Figure 21, inputs can be on the axon itself (A), the cell body (B) and on the dendrites (C and D). Some inputs are on dendritic membrane close to the cell body (C), others arrive on the tips (D) far away from the cell body. Remember from Section 3 that current spreads passively and attenuates rapidly with distance. What chance does a synaptic potential arriving on a remote part of the dendrite have of being able to effect a change in the membrane potential of the cell

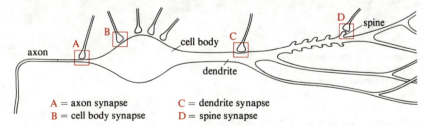

A = axon synapse   C = dendrite synapse
B = cell body synapse   D = spine synapse

FIGURE 21 Synaptic inputs can occur at varying distances from the cell body.

body? Synapses arriving on the cell body will obviously exert a stronger influence than others because of their immediate access to the axon hillock. Very often, these synapses are inhibitory, and no matter how many excitatory inputs are received on the dendrites an inhibitory input to the cell body blocks transmission. The position of synapses can therefore contribute to the integrative function of the neuron. When this is taken with the fluctuations in threshold potentials, the different diameters of dendrites, the timing of inputs and whether the inputs are excitatory or inhibitory, it is obvious that the neuron is a very sophisticated 'analyser' of information. In fact, the integration of information at synapses is very complex and outside the scope of this Unit. The point to realize is that impulses are not simply passed on through synapses, but decay, or are reinforced, depending upon activity in other circuits that also have inputs to those synapses.

5.3   Cells, circuits and systems

To explain the principles of communication within the nervous system, we have concentrated on the ways in which neurons connect with one another. The neurons we have considered are largely typical of the central nervous system (the brain and spinal cord) and peripheral sensory system. This has certainly provided us with a basis on which to build up a view of signalling, but what happens in a system such as that of our own brain which contains some 10^{12} neurons? As you can imagine, the millions of connections involved make the system very complex.

To show how neurons become arranged into more and more complex circuits we shall take a spinal cord reflex. The knee jerk reflex is something you have probably all experienced. When one leg is crossed over the other, the muscle that extends the knee joint is relaxed. A sharp tap on the tendon of the knee pulls the muscle, stretching it, and instantaneously the muscle contracts and the knee jerks.

Figure 22 illustrates the neuronal circuitry involved in this reflex. The tap on the knee stretches the muscle fibres, which stimulates a stretch receptor located deep within the muscle. The stretch receptor sends a stream of action potentials via the *afferent* (*sensory*) *nerve* to the spinal cord where an excitatory synapse conveys **afferent (sensory) nerve** the information to a large motor neuron. This initiates an action potential that proceeds to the muscle via an *efferent or motor nerve*, ending in a nerve–muscle **efferent (motor) nerve** junction that releases acetylcholine which causes muscular contraction. This circuit of neurons is called a *reflex arc*, and the knee jerk—a monosynaptic reflex—is **reflex arc** probably the simplest reflex of all.

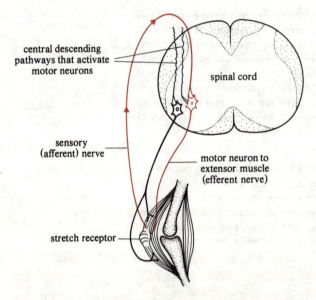

central descending pathways that activate motor neurons

spinal cord

sensory (afferent) nerve

motor neuron to extensor muscle (efferent nerve)

stretch receptor

FIGURE 22   Nerve circuits involved in the knee jerk reflex (a monosynaptic reflex). The reflex arc is shown in red. Note the possible involvement of central descending pathways from the brain (*in black*).

Now take a more involved system of neurons. Consider what happens when you touch a hot object. Muscles contract to remove the finger from the object, but often the body pulls back at the same time. The reflex action coordinates movement, this time in a number of muscles, which must involve more neurons and therefore more synapses. The circuit must include motor neurons at different places in the spinal cord because a number of sets of muscles are contracted. The other major part of the circuit must involve the brain because heat, and often pain, is experienced. So, the afferent sensory fibre must have access to relays that send information up the spinal cord to the brain and down the cord to various motor neurons. In this way, neural circuits are built up. The more complex the behaviour, the more neurons are brought into the circuit.

The complexity of nervous systems varies in different animal groups. In cnidarians, a true nervous system is present with neurons and synapses organized into a net of nerves. With most bilaterally symmetrical animals, neurons become concentrated into groups or *ganglia* (*sing.* ganglion) with a preponderance of ganglia in the head.

ganglia (*sing.* ganglion)

Invertebrate nervous systems are attractive to study because of their relatively small number of neurons, and because some of their axons (e.g. the squid giant axon) are large enough to take electrical recordings from easily.

In vertebrates, the nervous system can be divided into two parts (Figure 23).

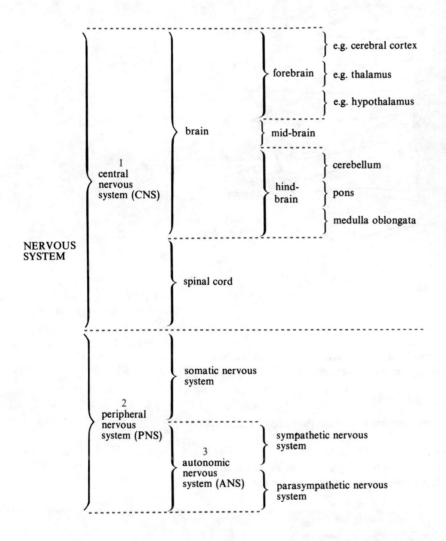

FIGURE 23   The organization of the human nervous system.

1   The *central nervous system* (*CNS*) comprises the brain and spinal cord.

central nervous system (CNS)*

2   The *peripheral nervous system* (*PNS*) includes the sensory receptors and their axons, together with motor fibres that innervate the main muscles of the body.

peripheral nervous system (PNS)*

In this Unit we are also concerned with:

3   The *autonomic nervous system* (*ANS*), which is concerned with the innervation of visceral organs and glands.

autonomic nervous system (ANS)*

The autonomic nervous system is well developed in mammals and birds and consists of two parts, separated both anatomically and functionally (see Figure 24 *overleaf*).

The nerve fibres of the *sympathetic* system arise from neurons in the trunk part of the spinal cord and communicate with neurons located in ganglia alongside the spinal cord. Efferent nerves from these ganglia innervate organs and glands at synapses that usually, but not always, use noradrenalin as the transmitter.

sympathetic and parasympathetic nervous systems*

The *parasympathetic* part of the autonomic nervous system arises from in front and behind the *sympathetic* system. These neurons use acetylcholine as the transmitter at the synapses to organs and glands.

27

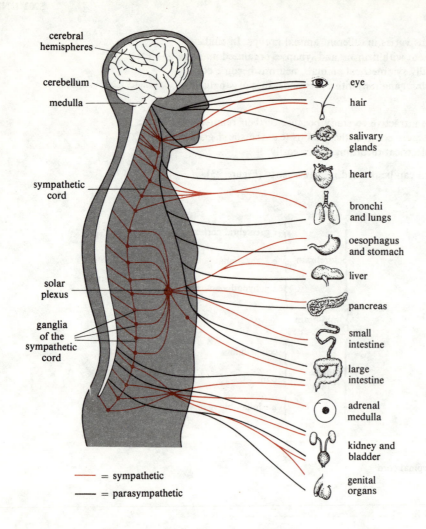

FIGURE 24 A schematic view of the autonomic nervous system in a man.

Labels on figure: cerebral hemispheres, cerebellum, medulla, sympathetic cord, solar plexus, ganglia of the sympathetic cord; eye, hair, salivary glands, heart, bronchi and lungs, oesophagus and stomach, liver, pancreas, small intestine, large intestine, adrenal medulla, kidney and bladder, genital organs.

— = sympathetic
— = parasympathetic

You may remember that both sympathetic and parasympathetic systems mostly innervate the same structures but often exert different effects[4].

☐ What effects does each have on the heart?

■ The parasympathetic vagus nerve slows the heart rate; the sympathetic system speeds it up.

The major control centres for the autonomic nervous system lie in the brain. These control regions (e.g. the hypothalamus; see Figure 25) are considered in

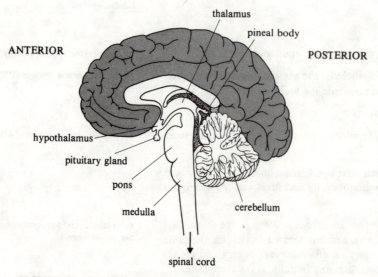

Labels on figure: thalamus, pineal body, ANTERIOR, POSTERIOR, hypothalamus, pituitary gland, pons, medulla, cerebellum, spinal cord.

FIGURE 25 A section through the human brain to show the major areas discussed in this Unit and other Units of this block, *Animal Physiology*. The cerebral cortex is shown as grey.

detail in Section 7 and in Unit 18, because it is in these areas that the various communication systems principally interact, and from these areas much of the endocrine system is regulated.

In the vertebrates, most of the ganglia are concentrated within a central structure, the brain. It is not possible to consider brain structure and function in detail here, but remember that the basis of functioning is exactly as we have described: signalling from neuron to neuron via synapses. The complexity of the brain is founded essentially in the millions of connections both within the brain and to other parts of the nervous system.

Summary of Section 5

Sensory receptors transduce 'environmental' stimuli into action potentials by generating local potentials. The resulting impulses are 'coded' according to their frequency, and receptors are specialized, interpreting stimuli and responding by switching on or off or changing the pattern of their responses. The integration of impulses occurs in the postsynaptic membrane, which acts as an analyser of inputs.

The knee jerk reflex is a monosynaptic pathway. More complex reflexes involve more neurons, and central control from the brain can dominate the expression of muscular activity (motor output). Neuronal circuits build up into nervous systems, reaching great complexity in mammals in which there is a division into central, peripheral and autonomic nervous systems. Control over these systems is predominantly exerted by the brain, whose functional basis lies in the transmission of impulses from neuron to neuron via synapses.

Objectives and SAQs for Section 5

Now that you have completed this Section, you should be able to:

★ describe ways in which the nervous system 'codes' the information it transmits.

★ demonstrate how the structure and function of synapses determine how inputs are integrated.

★ show how circuits can be 'engaged' in the nervous system with reference to simple and more complex reflexes.

To test your understanding of this Section, try the following SAQs.

SAQ 9 (*Objective 7*) Suggest three ways in which environmental information arriving at the sensory cell can be coded within the nervous system?

SAQ 10 (*Objective 8*) Construct a *simple* flow diagram to show where the stimulus is generally communicated within areas of the nervous system when you inadvertently prick your finger.

6 Hormonal control

The structure and function of the nervous system has been studied and discussed for several centuries. In contrast, most of what is known about hormonal control has been discovered during the past 80 years. Interest in hormones began with an experiment conducted by Arnold Berthold in 1848. He castrated six young cockerels (Figure 26) and then implanted one testicle into the body cavity of each of a number of the birds. The castrated birds with testicular implants exhibited the normal sexual behaviour for roosters and developed typical male features (i.e. large comb and wattles) whereas those without implants did not. Examination showed that the testicular implants had re-established a connection with the blood but not with the nervous system. Rather surprisingly, Berthold did not carry these experiments further, and their true significance was not realized until some 50 years later.

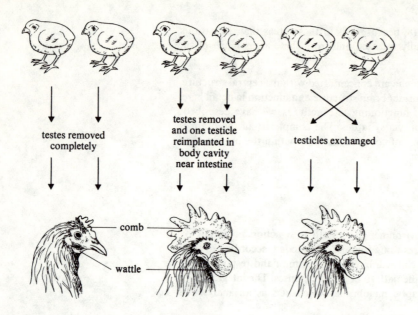

testes removed
completely

testes removed
and one testicle
reimplanted in
body cavity
near intestine

testicles exchanged

comb

wattle

FIGURE 26 Details of Berthold's
original experiment.

In a series of experiments between 1902 and 1905 two physiologists, Bayliss and
Starling, demonstrated unequivocally that coordination of the functions of organs
could occur without the intervention of the nervous system. They showed that a
chemical substance was secreted by cells in the small intestine (Figure 27) when

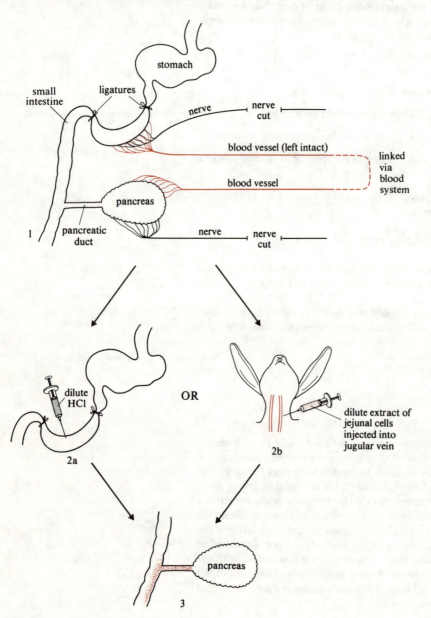

stomach

small
intestine

ligatures

nerve

nerve
cut

blood vessel (left intact)

linked
via
blood
system

blood vessel

pancreas

pancreatic
duct

nerve

nerve
cut

1

dilute
HCl

OR

dilute extract of
jejunal cells
injected into
jugular vein

2a

2b

pancreas

3

FIGURE 27 Bayliss' and Starling's
experiment. 1 A loop of small intestine
(jejunal region) in an anaesthetized
animal was tied at both ends and the
nerve supply was cut. 2a Dilute acid
was injected into the loop. OR 2b An
extract made from homogenized
intestinal cells was injected into the
jugular vein. 3 Either treatment resulted
in the secretion of pancreatic juice.

acidified food was emptied from the stomach. They named this chemical secretin and demonstrated that it was conveyed in the blood to the *pancreas*. Here, it stimulated some of the pancreatic cells to produce an alkaline juice that was then discharged through a duct into the intestine. Secretin is extremely potent and minute quantities are able to produce very marked effects. Starling suggested the name '*hormone*' (from the Greek for 'to arouse' or 'excite') as a general term to describe substances that are secreted into the bloodstream in minute amounts from one tissue and influence the activity of a distant target tissue. The technical term for the study of hormones is endocrinology.

pancreas*

hormones*

6.1 Hormones

You are not expected to remember the chemical structures of different hormones. This information is included in the text so that you can appreciate what is meant when several hormones are said to be similar or related. Neither are you expected to remember fine details such as the level of particular hormones in the blood. Concentrate instead on learning the general points that these details illustrate.

6.1.1 Hormone structure

Two main groups of hormones are known (Figure 28), those synthesized from fatty acid precursors and those produced from amino acids or closely related compounds.

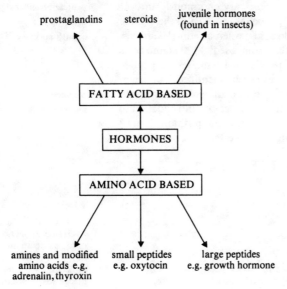

FIGURE 28 Various groups of hormones and their structural relationships. Most neurotransmitters are amines or modified amino acids. Prostaglandins are discussed in Unit 18.

Hormones based on *fatty acids* are relatively small in size and fairly constant in structure because different hormones are generated by small modifications of a basic molecule. For example, *steroid hormones* are all based on a molecular structure, consisting of four rings of carbon atoms, that is synthesized from cholesterol (Figure 29). In vertebrates, the important groups of steroids are the *oestrogens*

fatty acids

steroid hormones*

oestrogens

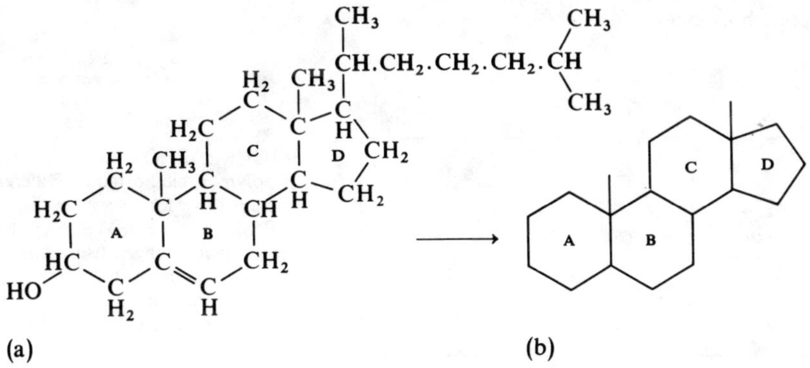

FIGURE 29 The relationship between (a) the parent cholesterol molecule and (b) the four-ringed steroid nucleus. All naturally occurring steroid hormones contain the 17 carbon atoms that make up the rings. Steroids differ from each other in terms of the number of carbon, hydrogen and oxygen atoms added onto the basic nucleus. (*You are not expected to remember these details.*)

(based on 18 carbon atoms—C_{18}), *androgens* (C_{19}), *progestins* and *corticosteroids* (C_{21}). Within each of these groups different hormones are synthesized by the addition (or removal) of oxygen and/or hydrogen atoms to (from) the basic

androgens
progestins
corticosteroids

steroid nucleus (Figure 30). Other hormones based on fatty acids are discussed in Unit 18 (prostaglandins) and the TV programme, *Insect Hormones*.

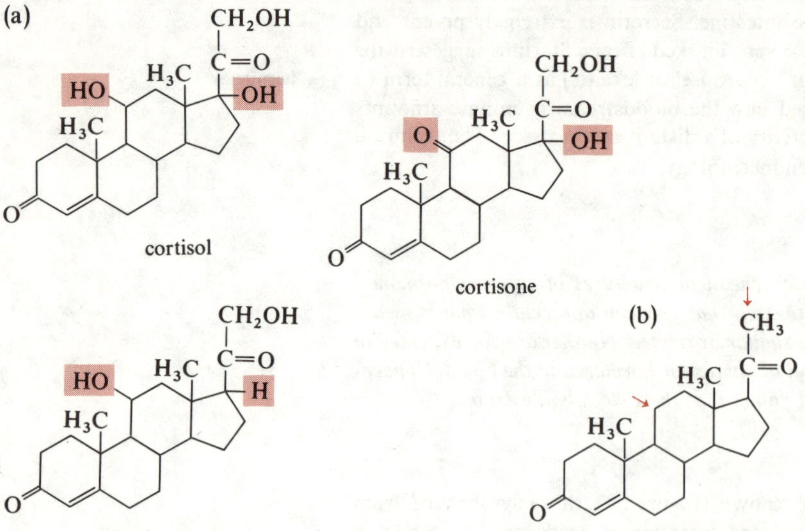

FIGURE 30 Some C_{21} steroids. Although the overall molecule is complex, the differences between the hormones are very simple. A group of corticosteroids is shown in (a). Different corticosteroids are synthesized by the addition (or removal) of hydrogen (or oxygen) atoms at key points on the molecule. (Compare the regions coloured in pink.) The progestins, for example progesterone (b), are also C_{21} steroids. However, their basic structure is different from that of corticosteroids. (Compare the positions indicated by red arrows in (b) with the same positions in (a). (*You are not expected to remember these structures.*)

In contrast, *peptide hormones*, based on amino acids, vary enormously in both size and structure. The hormone *thyroxin*, (T_4)*, which is involved in the control of tissue metabolism, is just two modified amino acids; whereas *growth hormone*, which is concerned with the regulation of growth, is composed of 190 amino acids and has a complex, three-dimensional structure. Distinct 'families' of peptide hormones containing similar amino-acid sequences can be identified. For example, *oxytocin* and *antidiuretic hormone (ADH)** are both small peptide hormones. These two hormones can produce quite different effects (see the Introduction to Units 16–25) but differ by only two amino acid substitutions, at positions 3 and 8 within the molecule (Figure 31). Oxytocin is found in all vertebrates, but different

peptide hormones*
thyroxin
growth hormone (GH)

oxytocin
antidiuretic hormone (ADH)

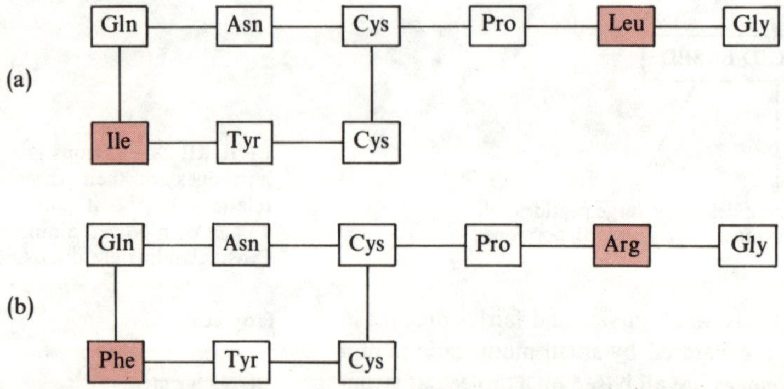

FIGURE 31 The amino acid sequences of (a) oxytocin and (b) antidiuretic hormone (ADH). The two cysteine molecules are joined by a disulphide bridge (S—S bond), so part of the molecule has a ring structure. The differences are shaded pink. (If you are unsure what the abbreviations stand for, refer back to Table 2, Unit 5.

vertebrate groups have slightly different versions of it (Figure 32). Such *polymorphism* (from the Greek for 'many forms') is quite common among the larger peptide hormones. Many neurotransmitters are essentially modified amino acids and are therefore related to the peptide hormones.

hormone polymorphism

	POSITION								
	1	2	3	4	5	6	7	8	9
OXYTOCIN (mammals)	Cys	Tyr	Ile	Gln	Asn	Cys	Pro	Leu	Gly
MESOTOCIN (birds, reptiles, amphibians)	Cys	Tyr	Ile	Gln	Asn	Cys	Pro	Ile	Gly
ISOTOCIN (bony fishes)	Cys	Tyr	Ile	Ser	Asn	Cys	Pro	Ile	Gly

FIGURE 32 Oxytocin is an example of a polymorphic hormone. Different groups of vertebrates have slightly different oxytocins. Mesotocin has one amino acid changed (isoleucine at position 8), and isotocin has two changes (serine at position 4 and isoleucine at position 8). The differences are shown in pink.

* Many hormones have long names that often indicate their function or origin. Hence, antidiuretic hormone *inhibits* the production of urine (Units 22 and 23). However, these are usually shortened to a set of initials for convenience: ADH = *anti*diuretic *h*ormone and T_4 = tetraiodothyronine = thyroxin.

32

6.1.2 The synthesis and storage of hormones

Hormones based on amino acids are usually stored in membrane-bound vesicles (100–500 nm in diameter) within the cell cytoplasm giving such endocrine cells a characteristic appearance.

☐ The hormone, adrenalin (relative molecular mass 183) is stored in membrane-bound vesicles that also contain Ca^{2+}, ATP and proteins. The various compounds interact with the hormone to form complexes with a relative molecular mass much larger than that of adrenalin. Can you suggest a possible rationale for this? (Recall the concept of osmotic pressure; Units 9 and 10, Section 2.1.1.)

■ If the vesicles simply contained large quantities of adrenalin, the osmotic pressure within the vesicles would be very high (osmotic pressure depends on the number of *freely moving particles* within a solution) and water would then enter the vesicles, until they eventually burst. The formation of aggregates of high relative molecular mass greatly reduces this problem.

Fatty acid hormones usually accumulate within the cell cytoplasm as clear lipid droplets, but they are sometimes stored in the form of lipid–protein aggregates, which may be bound in the membrane. Steroid-producing cells have a distinctive ultrastructure because of an abundance of smooth and rough endoplasmic reticulum and the presence of specialized mitochondria with tubular cristae. These mitochondria contain enzymes essential for steroid synthesis.

See The S202 Picture Book for a steroid-secreting cell.

Many hormones are synthesized in the form of inactive *parent molecules* (or *prohormones*) that subsequently become active after modification by enzymes. This post-synthetic processing may occur before or after secretion. Sometimes several different hormones may be synthesized as a result of enzymes cutting up the same parent molecule in different ways (Figure 33).

precursor (parent) molecule*
(e.g. prohormone)

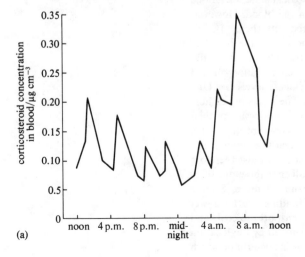

FIGURE 33 At least five different hormones can be produced from the same parent protein in certain endocrine cells of the pituitary gland. The hormones are cut out of the parent protein by specific enzymes. Some of these cells secrete several of these hormones, others secrete predominantly one hormone. The relationship of the hormones to the parent protein is indicated in the diagram.

6.1.3 The release of hormones and the duration of their action

The discharge of hormones from endocrine cells is controlled by nervous or hormonal or metabolic stimuli, which serve to alter the rate of release into the bloodstream. As with neurotransmitters, Ca^{2+} seems to play an important part in the secretion process. The release of hormones is not random, but is controlled in such a way that distinct rhythms or pulses of hormone release often occur (Figure 34). These may be circadian (i.e. they follow an approximate 24-hour cycle) or

FIGURE 34 (a) Changes in human corticosteroid levels over a 24-hour period. Levels are low at night and higher during the day. (b) Corticosteroid levels in the rat. Here levels are highest at night.

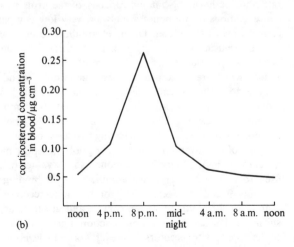

33

seasonal, but other periodicities are known (e.g. in the hormones controlling the human menstrual cycle; see Unit 18). However, some hormones are secreted more or less continually (e.g. thyroxin and triiodothyronine from the *thyroid gland*).

thyroid gland*

The *half-life of hormones* within the bloodstream varies from a few seconds to almost a week, depending on the hormone. The activity of the hormone is terminated by enzymes in the liver and target tissues that break down the hormone. In addition, small hormones may be lost as blood passes through the kidneys. The kidneys act as a selective filter; molecules with a relative molecular mass of 10 000 or less tend to pass into urine and are therefore excreted.

half-life of hormones*

Many small or insoluble hormones form complexes with large proteins found in the blood or produced by the parent endocrine cells. These *hormone-binding proteins in the plasma* perform several functions.

hormone-binding proteins in plasma*

1   They prevent the rapid excretion of the hormone when the blood is filtered by the kidney. (The hormone–protein complex is too large to pass through the filter.)

2   They prevent the rapid destruction of the hormone by enzymes.

3   They provide a source of hormone for slow release.

Slow release of hormone is possible because the amount of hormone bound to these plasma proteins is always in equilibrium with a small amount of '*free*' (unbound) *hormone*. Receptor molecules have a higher affinity for hormones than the plasma proteins do, so 'free' hormone tends to bind to receptors (when these are available; Figure 35). When this happens the equilibrium between protein-bound hormone in the plasma, 'free' hormone, and receptor-bound hormone

'free' hormone*

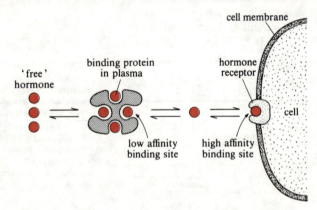

FIGURE 35   Equilibrium between 'free' hormone, protein-bound hormone in the plasma and receptor-bound hormone.

shifts, and hormone dissociates from the plasma protein. Because free hormone is being continually slowly broken down by enzymes the system never quite comes to equilibrium; instead, a steady trickle of hormone is released from the binding protein. Binding proteins in the plasma prolong the 'life' of small hormones.

6.1.4   Hormone concentrations

Figure 36 summarizes the various factors that govern hormone levels in the bloodstream. The *in vivo* (physiological) levels of particular hormones within the bloodstream vary widely and appear to be related to the volume of blood in the system (which is related to the animal's size), physiological state, metabolic rate and the hormone-binding capacity of the proteins in the blood. However, concentrations of hormones are always very low compared with those of other constituents of the blood. Levels of antidiuretic hormone within the blood of the rat are usually in the region of 10^{-11} mol l^{-1} but increase by about 25-fold during dehydration to 2.5×10^{-10} mol l^{-1}. Hormones such as glucagon and insulin, which are concerned with the regulation of blood glucose levels (see Unit 17) are normally present at levels of about 10^{-9} mol l^{-1}. The levels of sex hormones vary widely between 10^{-7} mol l^{-1} and 10^{-10} mol l^{-1} as a result of their rhythmic release (see Unit 18). Detecting a 10^{-10} mol l^{-1} concentration of a hormone is about equivalent to tasting a single spoonful of sugar in a cup of coffee the size of a conventional swimming pool! Receptor molecules must therefore have a very high affinity for hormones, so K_d values (recall that these are often used to compare the relative affinities of two binding sites; Unit 6, Section 3; Units 9 and 10, Section 2.2) for hormone–receptor binding are far lower than those for binding between enzyme and substrate or transport molecule and substrate. The site of a hormone's action is generally determined by the location of the appropriate receptors. However, hormone levels may also determine which target tissues are influenced—some target cells are more sensitive than others to a particular hormone. (Note that this is comparable to the situation sometimes encountered in neuronal networks where different neurons may have different thresholds even though the same neurotransmitter is being used to trigger an action potential.)

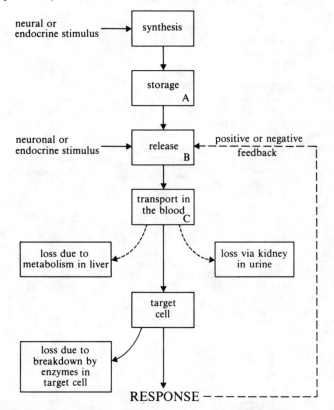

FIGURE 36 The life of a hormone. A *Storage* The hormone may be stored bound to a larger polypeptide, or be present as part of a larger parent polypeptide. This probably keeps osmotic pressure in the vesicle low. B *Release* The hormone may be released intermittently, continually, or in distinct pulses. C *Transport in the blood* The hormone may be bound to proteins in the plasma. This prevents loss in the kidney or liver. The location of the gland in relation to the blood system may also be important.

The role of feedback will be discussed in Section 7.2.

It is difficult to measure hormone levels because of the extremely low concentrations involved and the presence of binding proteins. Two different methods are generally used, *bioassay* and *radioimmunoassay* (see Box 1). Low concentrations also make the isolation of sufficient quantities of a hormone for chemical analysis something of a Herculean task (e.g. over 4 tons of sows' ovaries were processed to produce the 12 mg of crystals that were eventually used to elucidate the structure of the female sex hormone, oestradiol).

bioassay*
radioimmunoassay*
Box 1 is on p. 36.

If abnormally large concentrations of a hormone are injected into an animal, effects may be observed that are unrelated to the hormone's normal action. For example, the hormone oxytocin is normally involved in the control of milk ejection in mammals. However, if the concentration of oxytocin in a pig's blood is raised to $10^{-6}\,\text{mol}\,\text{l}^{-1}$, the output of urine is reduced.

☐ Urine output is normally regulated by antidiuretic hormone (ADH), and an injection of $10^{-9}\,\text{mol}\,\text{l}^{-1}$ ADH has the same effect on the production of urine as $10^{-6}\,\text{mol}\,\text{l}^{-1}$ oxytocin. Suggest a possible reason for this overlap in the action of the two hormones. (Recall Figure 31 and what you learned about the uptake of glucose and related sugars by the same transport molecule in Units 9 and 10, Section 2.2.)

■ Oxytocin and ADH are similar in structure, and structurally similar sugars will bind to the same transport molecules; but they bind with different affinities (i.e. the K_d value is different; Table 3, Units 9 and 10). High concentrations of oxytocin mimic the effects of ADH, and this suggests that the ADH receptor will also bind oxytocin (by analogy with the sugar situation). The need for high concentrations of oxytocin indicates that the ADH receptor has a lower affinity (a higher K_d value) for oxytocin.

This *cross-reactivity* between the two hormones does not normally occur because concentrations of oxytocin do not usually rise above $10^{-10}\,\text{mol}\,\text{l}^{-1}$. The responses produced by supranormal hormone concentrations are said to be *pharmacological* to distinguish them from *physiological* effects produced by normal (*in vivo*) concentrations. Cross-reactivity is important in medicine (e.g. when an endocrine gland develops a tumour and secretes excessive amounts of hormone, or when injections are used to correct a hormone deficiency; see Unit 17).

cross-reactivity of hormones

BOX 1   Methods of measuring hormone concentrations

1   Bioassay

This technique makes use of the effects produced by a particular hormone. For example, growth hormone (GH) stimulates the growth of cartilage in young mice. The thickness of the cartilage produced is proportional to the quantity of hormone present (Figure 37a). This response can be used as a basis for measuring levels of human growth hormone. By adopting an arbitrary standard (e.g. 1 mouse unit of GH produces a tibial cartilage thickness of 1 mm in 17 days; 2 mouse units produce 2 mm, etc.), it is possible to quantify unknown amounts of hormone in terms of these standard units. This is done by injecting the mice with an unknown quantity of hormone isolated from a standard sized sample of human blood (e.g. 10 cm^3) and then measuring cartilage growth after the set time. Alternatively, if known amounts of pure hormone can be assayed, in addition to the unknown blood sample, it is possible to plot a graph of cartilage thickness against the concentration of growth hormone (Figure 37b). Unknown concentrations of growth hormone can then be estimated using this standard graph. Bioassays can be devised for most hormones but they may not be particularly accurate because individual animals may vary in their responsiveness to the same quantity of hormone.

2   Radioimmunoassay

This technique makes use of (a) *antibodies* produced by the vertebrate *immune system* and (b) the fact that different vertebrates often use slightly different versions of the same hormone (see Figure 32). If human *prolactin* (*PRL*) is injected into a rabbit, its white blood cells classify this hormone as 'foreign' (the amino acid sequence of human PRL is different from that of rabbit PRL) and try to inactivate the human prolactin by secreting large quantities of antibody molecules that are highly specific for human prolactin.

The assay is in two stages. The first involves the construction of a standard curve by incubating different known quantities of hormone with the same *fixed* number of antibody molecules (Figure 38a, step 1). If the number of antibody molecules always exceeds the number of hormone molecules, the number of antibody binding sites left unoccupied (Figure 38a, step 2) after incubation is inversely proportional to the amount of hormone added (the more hormone, the fewer unoccupied sites). The number of unoccupied sites is estimated by adding a *fixed* quantity of *radioactively labelled* hormone to the original hormone–antibody mixture (Figure 38a, step 3). Some of the radioactively labelled hormone fills up the unoccupied binding sites and the rest is left in solution as 'free' (unbound) radioactive hormone. The test-tube is then centrifuged to remove the antibody molecules leaving the 'free' radioactive hormone in the liquid fraction (supernatant). The amount of radioactive hormone bound to the antibodies can then be measured and expressed as a percentage of the quantity of radioactive hormone originally added. The percentage of radioactive hormone that is bound is a measure of the number of unoccupied antibody binding-sites and is thus inversely proportional to the amount of non-radioactive hormone in the original sample. This technique is

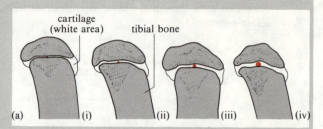

FIGURE 37a   The effect of growth hormone on the width (*red spot*) of cartilage produced on the tibia of mice 17 days after injection. Mice were injected with: (i) a control-saline injection containing no hormone; (ii) 1 µg of growth hormone; (iii) 5 µg of growth hormone; (iv) 20 µg of growth hormone.

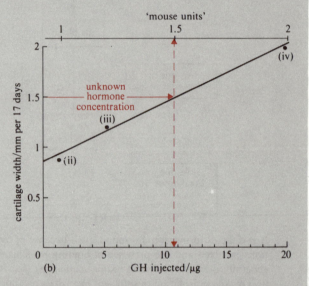

FIGURE 37b   Standard graph showing the dose of growth hormone (measured in arbitrary units *or* µg GH injected) against response (cartilage growth). Such a plot is also known as a 'dose–response' curve. An unknown concentration of hormone produces cartilage 1.5 mm wide in 17 days; therefore the amount of unknown hormone must be about 10.8 µg of growth hormone *or* 1.5 'mouse units', if an arbitrary scale is being used. (Note that the relationship between dose and response is usually more complex (curvilinear rather than linear) if a large range of doses is studied. But this can usually be allowed for; see Figure 38.)

FIGURE 38a   The radioimmunoassay technique.

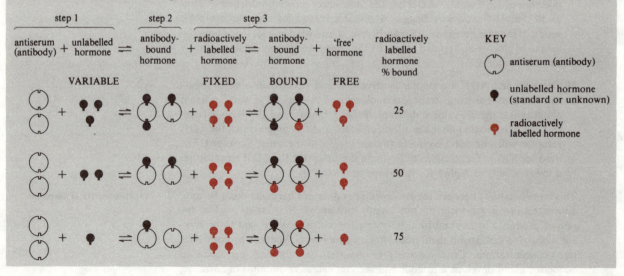

used to construct a standard curve (Figure 38b), which can then be used to relate the percentage of bound radioactive hormone obtained from samples containing unlabelled hormone to absolute hormone concentrations (e.g. in Figure 38b, 43 per cent of radioactive hormone bound = a prolactin concentration of $1.25 \, \text{ng cm}^{-3}$). Provided a suitably specific antibody can be produced and isolated, radioimmunoassay is more accurate and reliable than bioassay and can be used to measure the concentrations of substances other than hormones. Its application has had such an enormous impact on medical diagnosis that its originator Rosalind Yalow was awarded a Nobel Prize for Physiology or Medicine in 1977.

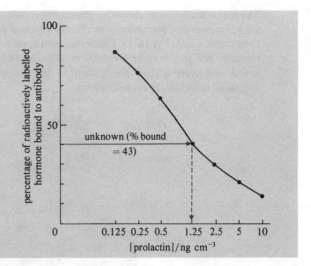

FIGURE 38b A standard graph showing the percentage of radioactively labelled hormone bound to antibody against the hormone concentration (measured in ng cm^{-3}). If the x axis is plotted on a linear scale, a curve is obtained. It is not very easy to read off unknown sample concentrations from such a graph, so it is more usual to plot hormone concentrations on a log scale, which produces a relatively straight line.

Summary of Section 6.1

1 Hormones can be broadly subdivided into those based on (a) fatty acids and (b) amino acids. Steroids are important examples of the first group and polypeptide hormones of the second. Many neurotransmitters are modified amino acids and therefore related chemically to the peptide hormones.

2 The amino acid sequence of a particular peptide hormone may vary slightly in different animal species.

3 Many hormones are synthesized as inactive parent molecules which are converted to the active form either before or after leaving the endocrine cell.

4 Distinct families of hormones, with short sequences of acids in common, can be identified. For some hormones this can be explained on the basis of each one being derived by post-synthetic processing from a common parent molecule.

5 Hormone-producing cells have recognizable ultrastructural features that relate to the type of hormone they produce.

6 Hormones may be released (a) more or less continuously, (b) in short distinct pulses (say, over a 24 hour period) or (c) in longer cycles.

7 The duration of a hormone's action depends on the longevity of the hormone. This is governed by (a) its rate of breakdown in the liver and at the target tissue and (b) its rate of excretion, which is related to its size. The effective life of small hormones is often extended by interactions with binding proteins that are present in the blood.

8 Hormone receptors have very high affinities (low K_d values) because concentrations of hormones are normally relatively low.

9 Hormone levels are difficult to measure because of the low concentrations; bioassay and radioimmunoassay are the usual methods used.

10 At supranormal concentrations structurally related hormones may cross-react with each other's receptor molecules. This phenomenon is important in certain medical situations.

6.2 Endocrine glands

The cells that produce hormones are often gathered together forming distinct tissues termed *endocrine glands*. In vertebrates, these organs are richly supplied with networks of small blood vessels (capillaries) that facilitate the secretion of hormones directly into the bloodstream. Endocrine tissues are often referred to as 'ductless' glands to distinguish them from other 'glandular' tissues such as sweat glands and salivary glands. The latter are termed *exocrine glands* and release their fluid secretions via *ducts*. In contrast to endocrine glands (from the Greek prefix *endo* = within), exocrine glands (from the Latin prefix *exo* = out of) discharge their secretions to the *ex*terior (or what is effectively the exterior, i.e. the digestive tract) of the organism.

endocrine glands*

exocrine glands

The major mammalian endocrine glands are shown in Figure 39. Information on the hormones they secrete and the effects that these produce is summarized in the Introduction to Units 16–25. Endocrine glands seem to be less common in invertebrates, but this may be a reflection of present-day ignorance. Fairly conventional endocrine glands are known to be present in cephalopod molluscs, crustaceans, echinoderms and insects.

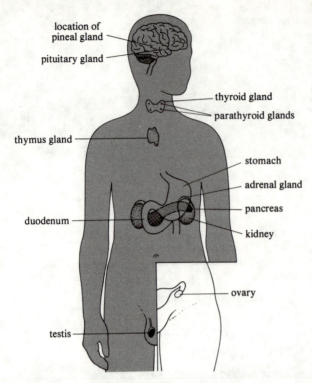

location of pineal gland
pituitary gland
thyroid gland
parathyroid glands
thymus gland
stomach
adrenal gland
pancreas
duodenum
kidney
ovary
testis

FIGURE 39 Major sites of hormone production in mammals.

Suspected endocrine glands are often identified by experiments involving the *ablation* (removal) of the gland, followed by *reimplantation* (or an injection of hormone). Removal of the putative gland should generate deficiency symptoms. Reimplantation of the missing gland ought to relieve these symptoms if the blood supply to the gland is resumed. (Note the similarity of this with Berthold's original experiments.) The injection of extracts containing the 'active' product (hormone) secreted by the gland should also relieve symptoms of deficiency. These methods have been used extensively, but they are not always appropriate, for the following reasons.

ablation and reimplantation experiments

1   Some endocrine cells are found scattered among other cells that form part of a larger organ. For example, the pancreas (see Unit 17) is a mixture of endocrine and exocrine glands. Similarly, the testis contains both sperm-producing cells and endocrine cells (see Unit 18).

2   Some endocrine tissues contain a variety of cell types producing different hormones. For example, the mammalian *adrenal gland* (Figure 40) contains two

adrenal gland*

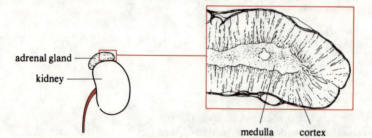

adrenal gland
kidney
medulla   cortex

FIGURE 40   The location and structure of the mammalian adrenal gland. The enlargement is a drawing of a low-power histological section. Note the single large vein in the centre of the medulla.

noradrenalin
adrenalin

distinct hormone-producing regions: an inner medulla that secretes *noradrenalin* and *adrenalin*, and an outer cortex that secretes mainly corticosteroids. The cortex can be further subdivided into three distinct regions each of which produces different C_{21} steroid hormones.

3   A gland which serves a single function may have different anatomical appearances in different groups of organisms. For example, the adrenal cortex and medulla are two quite separate structures in certain species of fish.

4   Any surgical operation will obviously affect a whole range of physiological processes for a short while. However, this can be allowed for to some extent by doing control experiments in which sham operations are performed; for example, a piece of muscle may be implanted rather than endocrine gland.

The removal–reimplantation or removal–replacement approach to identifying endocrine glands is therefore not ideal and can give very misleading results. As you might imagine, the removal of a complex, multifunctional organ produces a wide range of physiological changes that are not necessarily related to particular hormonal effects. Consequently, modern endocrinologists often make more use of ultrastructural and biochemical features that are known to be unique to endocrine cells as a means of identification. Many vertebrate endocrine glands contain complex blood systems that permit short-distance communication between different regions of the gland. Let us take an example. The relative amounts of noradrenalin and adrenalin secreted by the adrenal medulla are controlled by corticosteroids. Corticosteroids stimulate the cells of the medulla to produce an enzyme that converts noradrenalin to adrenalin. Corticosteroids thus increase the amount of adrenalin secreted. This particular effect is important in modifying the organism's response to prolonged stress. However, the cells of the medulla respond only to high levels of corticosteroids—levels far higher than those found in the main bloodstream. How are these levels achieved? The gland is fed by several arteries, but all the blood leaving the gland drains into a single vein that runs through the centre of the medulla. Consequently, venous blood from the cortex has to pass through the medulla before it joins the main blood supply where the concentration of corticosteroids becomes diluted.

Objectives and SAQs for Section 6

Now that you have completed this Section, you should be able to:

★ outline the various factors that govern the 'effective life' of a hormone and hence the duration of its action.

★ explain why the concentration of a hormone is important in determining its target tissue.

★ on the basis of its ultrastructure, distinguish a cell that produces peptide hormones from one that produces steroid hormones.

★ explain why the removal–reimplantation technique practised by Berthold is of limited value in the identification of some endocrine glands.

To test your understanding of this Section, try the following SAQs.

SAQ 11   (*Objectives 9 and 10*)   Are the following statements *true* or *false*? Give reasons for your answers.

(a) Neurotransmitter molecules and peptide hormones are more closely related chemically than steroid hormones and peptide hormones, or carbohydrates and peptide hormones.

(b) Cross-reactivity is more likely to occur between progesterone and cortisol than between growth hormone and cortisol.

(c) Small hormones bind to hormone-receptor molecules only.

(d) The enzyme lactic dehydrogenase (LDH) has the same amino acid sequence irrespective of the animal or tissue from which it is isolated. Similarly, growth hormone isolated from different animals should also have identical sequences of amino acids. This is because the function of a protein depends on its entire amino acid sequence.

(e) Steroid-secreting cells have specialized mitochondria and a prominent, smooth endoplasmic reticulum.

(f) Some endocrine cells secrete inactive forms of hormones that are later modified by enzymes in the blood or at the target tissue.

SAQ 12   (*Objectives 9 and 10*)   List *four* factors that determine which target tissue responds to a particular hormone and the duration of the response.

SAQ 13 (*Objectives 9 and 10*) Doctors often ask hospital laboratories to analyse blood and/or urine samples for specific hormones in order to diagnose certain diseases.

(a) Explain why the doctor needs to know the approximate time of day the sample was taken in order to interpret the information provided by the laboratory.

(b) What sort of hormones are likely to appear in the urine?

(c) What kind of information could this analysis yield?

SAQ 14 (*Objectives 9 and 10*) Consider the following facts. The hormone insulin contains amino acid sequences that are similar to those of certain other hormones. Some diabetics produce insufficient amounts of insulin. Insulin can be isolated from various animals, but each type of insulin has a slightly different sequence of amino acids. Insulin deficiency in humans can be corrected by regular injection of solutions containing animal insulins.

(a) Suggest one reason why the amount (concentration) of insulin injected has to be carefully controlled and kept as low as possible.

(b) Suggest one reason why insulins of non-human origin may become less effective over a period of time. (Recall how antibodies are prepared for radio-immunoassays.)

SAQ 15 (*Objectives 9 and 10*) The hormone thyroxin has a half-life in the bloodstream of about a week. In certain diseases the activity of the thyroid gland declines, resulting in a decrease in the secretion of the hormone thyroxin. This decline can be detected by radioimmunoassay. Suggest why blood samples that are to be analysed for thyroxin are usually either heated slightly or incubated briefly with a solution of relatively high pH before antibodies to thyroxin are added. To answer this question, recall (a) the size of thyroxin, (b) how hormone secretion is prevented and (c) the effect of heat or pH on protein structure.

SAQ 16 (*Objectives 9 and 10*) The pancreas is known to secrete the peptide hormone glucagon. The removal of the pancreas causes a marked rise in the level of blood glucose.

(a) Would it be correct to deduce that glucagon normally depresses the level of blood glucose? Give reasons for your answer.

(b) How might you test the validity of this deduction?

7 Neurosecretion

> It appears indeed as if there is a much closer relationship between nervous and glandular (endocrine) tissues than is commonly recognized.
>
> (Comment made in a review by Ernst and Berta Scharrer published in 1940)
>
> We have just heard some very interesting things—and also a great deal of nonsense!
>
> (Comment addressed to Ernst Scharrer when he presented the concept of neurosecretion at a scientific conference in London in 1953)

During the 1920s a German scientist, Ernst Scharrer, noticed that certain neurons in the hypothalamic region of the brain of the minnow (*Phoxinus phoxinus*) had structural features similar to those of endocrine cells. The axons of these neurons passed out of the *hypothalamus* and into the pituitary gland. Instead of making synaptic contact with other nerves or effector (target) cells, these axons ended in large bulbous swellings that were closely applied to certain of the numerous capillaries that pass through the pituitary gland. These findings led Scharrer to suggest that at least some of the hormones released from the pituitary gland were actually synthesized in neurons. If this were so, the pituitary might act merely as a storage and release depot rather than as a conventional endocrine gland. At the time, the idea was almost universally rejected by the scientific community because it cut right across the concept of two distinct communication systems. However, Ernst Scharrer and his wife Berta persisted in their studies, he looking for similar phenomena in other vertebrates while she searched among the invertebrates. By the early 1950s a considerable body of evidence supported Scharrer's original hypothesis, and hormone-secreting nerves had been identified in a number of

hypothalamus*

invertebrates. By this time the phenomenon was becoming accepted and was termed *neurosecretion* to indicate that the nerves secreted material into the general bloodstream in much the same way as other endocrine cells. The secretory products were termed *neurohormones* to indicate their origin in the nervous system, and the organs involved in storage and release were termed *neurohaemal organs* because of the close association of nerve endings and blood capillaries within these structures.

In general, neurosecretory neurons differ from conventional neurons in the position and structure of the nerve endings, and in the size of granules found within the cell. Neurosecretory granules are much larger (200–500 nm in diameter) than those found at standard nerve synapses (usually 40–100 nm in diameter) and similar in size to those found in many endocrine cells. In comparison with conventional nerves, neurosecretory neurons have more nerve endings (Figure 41), which

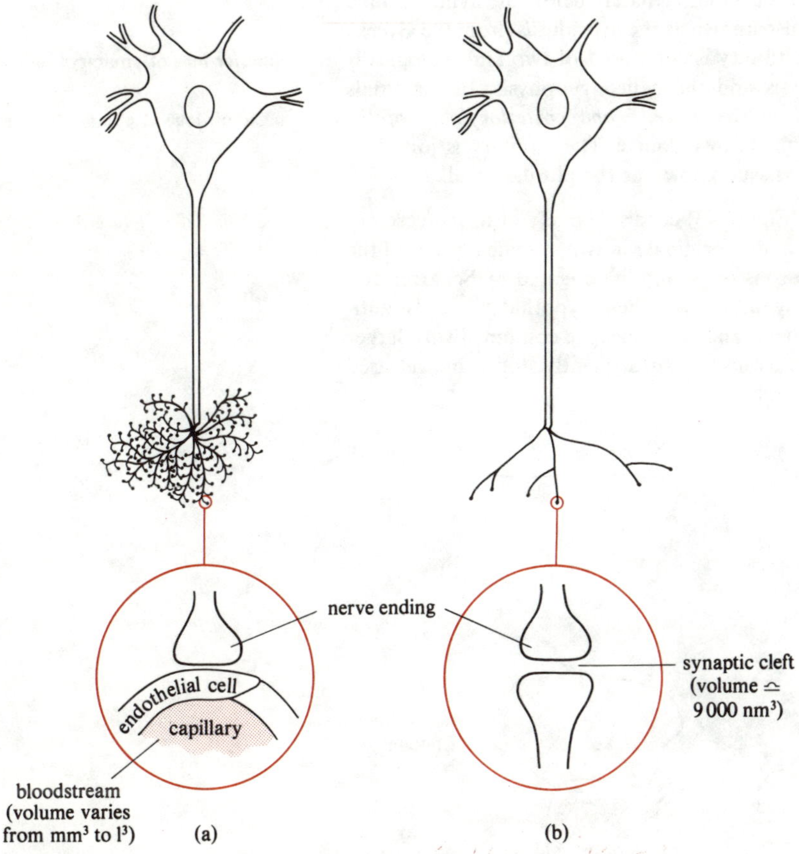

neurosecretion*

neurohormones*
neurohaemal organs

FIGURE 41 (a) A typical neurosecretory neuron. (b) A conventional neuron.

enables them to secrete relatively large quantities of neurohormones, once stimulated. This is an important structural adaptation because for these cells the 'synaptic cleft' (i.e. the blood supply) has an enormous volume compared with that of a conventional nerve. Like those of conventional neurons, neurosecretory axons produce and conduct action potentials but changes in potential develop more slowly and last longer. This seems to be another adaptation that increases the quantity of neurohormone secreted per stimulation.

7.1 The pituitary and the hypothalamus

This is a difficult but important Section. Figures 46 and 48 summarize most of the key information, and you should try to remember these. We discuss the idea of feedback and you may find it helpful to refer to the Foundation Course if you have difficulty understanding this concept.

The concept of neurosecretion provided new insights into the nervous system, and it also focused attention on the pituitary gland which, by this time, was known to be involved somehow in the regulation of various physiological processes. The *pituitary gland* is a small organ (the human pituitary weighs about 0.5–0.8 g) situated in the floor of the skull just above the roof of the mouth (Figure 42). The gland was discovered by the sixteenth century anatomist Vesalius, who misguidedly thought it was responsible for the secretion of pituita (nasal fluid), hence

pituitary gland (hypophysis)*

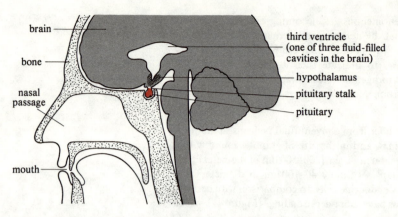

FIGURE 42   The location of the pituitary gland in humans.

the name pituitary. Because of its position immediately below the hypothalamic region of the brain, the pituitary is also known as the *hypophysis* (from the Greek: *hypo*, under; *physis*, growth). The pituitary is composed of two embryologically distinct tissues: the adenohypophysis and the neurohypophysis. In mammals these two sections of the gland form the *anterior and posterior lobes of the pituitary*, and we shall use these terms in this Course. The pituitary is joined to the brain by a thin stalk of nervous tissue known as the pituitary stalk.

anterior lobe of pituitary* (adenohypophysis)

posterior lobe of pituitary* (neurohypophysis)

The posterior lobe of the pituitary (Figure 43) consists largely of neurosecretory neuron endings, most of whose cell bodies originate in two specific regions of the hypothalamus. (These were the neurons originally investigated by Scharrer and his associates.) Two hormones are synthesized in these hypothalamic cells: anti-diuretic hormone (ADH) and oxytocin. The hormones are combined with larger proteins, then transported along the axons and subsequently stored and released from nerve endings in the posterior lobe.

See *The S202 Picture Book* for posterior pituitary.

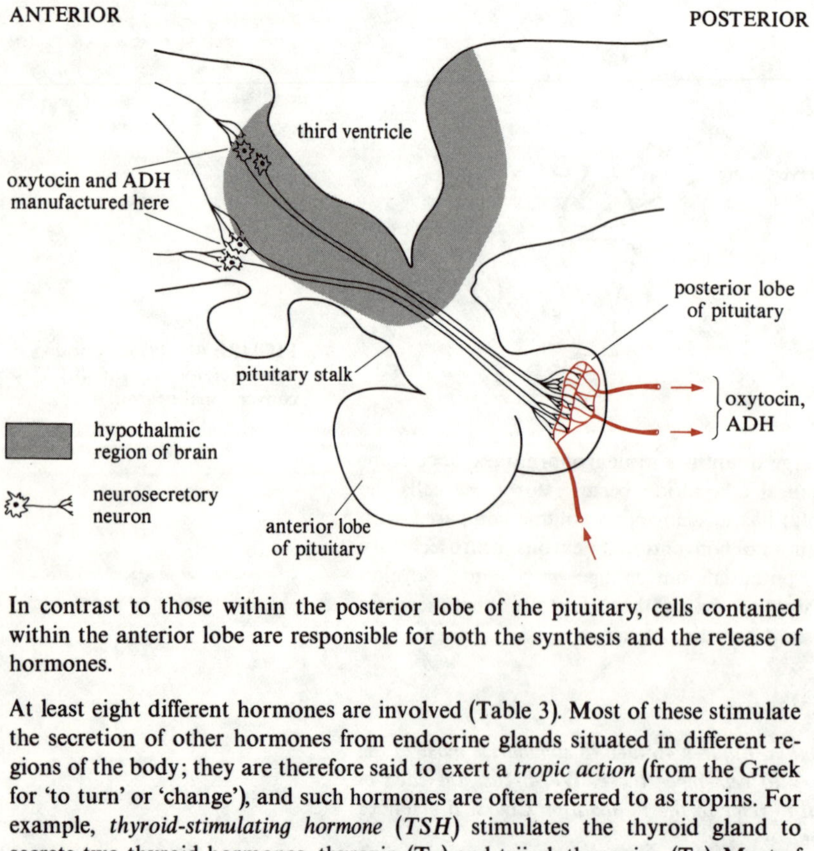

ANTERIOR   POSTERIOR

third ventricle

oxytocin and ADH
manufactured here

posterior lobe
of pituitary

pituitary stalk

oxytocin,
ADH

hypothalmic
region of brain

neurosecretory
neuron

anterior lobe
of pituitary

FIGURE 43   The posterior lobe of the pituitary and its associated neurosecretory neurons in the hypothalamus.

In contrast to those within the posterior lobe of the pituitary, cells contained within the anterior lobe are responsible for both the synthesis and the release of hormones.

See *The S202 Picture Book* for anterior pituitary.

At least eight different hormones are involved (Table 3). Most of these stimulate the secretion of other hormones from endocrine glands situated in different regions of the body; they are therefore said to exert a *tropic action* (from the Greek for 'to turn' or 'change'), and such hormones are often referred to as tropins. For example, *thyroid-stimulating hormone (TSH)* stimulates the thyroid gland to secrete two thyroid hormones, thyroxin (T_4) and triiodothyronine (T_3). Most of these hormones are relatively large peptides. Unlike the posterior lobe of the pituitary, the anterior lobe did not appear to have a nerve supply, so it was rather a surprise when experiments showed that electrical stimulation of certain parts of the hypothalamus dramatically increased the secretion of some anterior lobe hormones and decreased the secretion of others. A possible solution to this problem became evident when the blood system of the pituitary was investigated (Figure 44).

tropic action*

thyroid-stimulating hormone (TSH)*

TABLE 3 Hormones produced by the anterior lobe of the pituitary (adenohypophysis)

Hormone (name used in this Course)	Abbreviation	Target cells	Action
thyroid-stimulating hormone	TSH	thyroid gland	stimulates the synthesis and secretion of thyroxin and triiodothyronine
follicle-stimulating hormone	FSH } these two hormones are collectively termed *gonadotropins*	testis and ovary	controls the development and maturation of germ cells, i.e. spermatozoa and ova
luteinizing hormone	LH }	testis and ovary	controls the secretion of the steroid hormones responsible for the male and female sexual characteristics; also triggers ovulation
prolactin	PRL	mammary gland	stimulates milk production
		corpus luteum	maintains the secretion of oestrogen and progesterone
		liver	stimulates the production of a pheromone that controls maternal behaviour in some mammals
		fish gills	involved in the maintenance of salt balance and in osmoregulation
growth hormone	GH	liver	stimulates the liver to form somatomedins, which alter the metabolism of tissues (e.g. liver, muscle, adipose tissue)
adrenocorticotropic hormone	ACTH	adrenal cortex	stimulates the synthesis and release of glucocorticoids
β-lipotropin	β-LPH	adipose tissue	mobilizes lipids; but may be more important as a precursor (parent molecule) for endorphin and enkephalins
β-endorphin, enkephalins		neurons	alter the activity of certain neurons
melanocyte-stimulating hormone	MSH	melanocytes* (also neurons)	darkens the skin in lower vertebrates; controls hair colouration in some mammals; affects the activity of some neurons

* Cells responsible for skin colour change. Other terms will be explained in later Units.

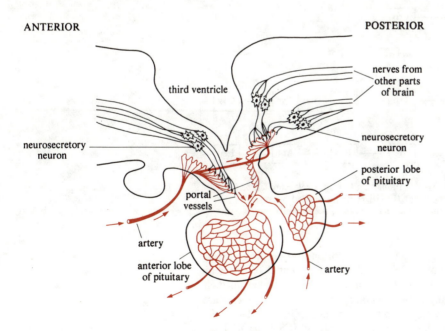

ANTERIOR

POSTERIOR

third ventricle

nerves from other parts of brain

neurosecretory neuron

neurosecretory neuron

portal vessels

posterior lobe of pituitary

artery

anterior lobe of pituitary

artery

FIGURE 44 The blood supply to and from the pituitary gland.

□ Suggest how the hypothalamus could influence the release of hormones from the anterior lobe of the pituitary.

■ You should have noticed from Figure 44, that (a) there is an elaborate capillary network that is arranged so that blood passing into the hypothalamus drains into the anterior lobe and (b) a number of neurosecretory nerve endings are closely associated with this capillary network. Given this anatomical arrangement, you could postulate (by analogy with what you know about the adrenal gland; Section 6.2) that the neurohormones released into the bloodstream from hypothalamic neurosecretory nerves after electrical stimulation either stimulate or inhibit the secretion from the target cells in the anterior lobe.

☐ Suggest a method of testing your hypothesis.

■ One way would be to disrupt (i.e. cut) the blood capillaries (or portal vessels as they are termed) that connect the hypothalamus and the anterior lobe of the pituitary and then repeat the electrical stimulation experiments. Stimulating the appropriate regions of the hypothalamus should now have no effect on the secretion of hormones from the anterior lobe.

This hypothesis was originally put forward by Geoffrey Harris at Oxford University in the 1940s. Although his ideas have turned out to be correct, they were not easy to substantiate. To give the hypothesis credibility, the neurohormones (or *releasing and release-inhibiting factors*, as these are now called) had first to be isolated. They proved to be very elusive, so much so that many physiologists began to think that the hypothesis was wrong.

releasing and release-inhibiting factors (neurohormones)*

However, about 30 years later, in 1969, two different research groups announced (almost simultaneously!) that they had isolated and chemically identified a releasing factor. This was *thyroid-stimulating hormone releasing factor* (TRF), which stimulates certain cells of the anterior lobe of the pituitary to release thyroid-stimulating hormone. This was a major milestone in endocrinology and the leaders of the two groups, Roger Guillemin and Andrew Schally, were subsequently awarded Nobel Prizes for Physiology or Medicine in 1977. Some idea of the task involved can be gauged from Guillemin's estimate that, weight for weight, the cost of isolating the first milligram of thyroid-stimulating hormone releasing factor was two or three times more than the cost of a kilogram of the moonrock brought back by the Apollo 13 spacecraft. His group alone processed almost 5 million hypothalamic fragments from 500 tons of sheep brain over a period of 4 years!

thyroid-stimulating hormone releasing factor (TRF)

Several other releasing and release-inhibiting factors have now been isolated and, like oxytocin and antidiuretic hormone (Figure 31), they are all relatively simple peptides (Figure 45). This has practical implications because chemists can manufacture small peptides relatively cheaply whereas the chemical synthesis of the

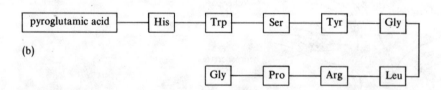

(a)

(b)

FIGURE 45 The structure of (a) thyroid-stimulating hormone releasing factor (TRF) and (b) luteinizing hormone-releasing factor (FSH/LH-RF), so-called because this releasing factor is also thought to control the secretion of follicle-stimulating hormone (FSH).

large peptide hormones secreted by cells of the anterior pituitary is prohibitively expensive. Various drug companies quickly realized the potential of synthetic versions of these factors, which is why so much effort was expended on their isolation. Synthetic versions of thyroid-stimulating hormone releasing factor are currently used in a very sensitive test for pituitary malfunction, and chemically modified versions of luteinizing hormone-releasing factor (FSH/LH-RF) are being tested as contraceptives (see Unit 18).

Figure 46 summarizes the interrelations of the hypothalamus and various regions of the pituitary. As you can see, the pituitary is involved in the control of a wide range of organs and physiological processes.

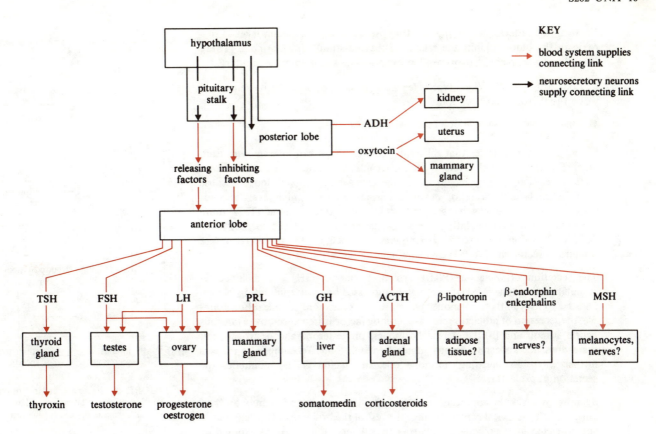

KEY

→ blood system supplies
connecting link

→ neurosecretory neurons
supply connecting link

FIGURE 46 A summary of the releasing factors and stimulating hormones from the hypothalamus and the pituitary.

The release of many pituitary hormones is a carefully regulated process involving feedback loops[5] that are similar to those discussed in Unit 7 (Section 4). If two or more variables are interdependent, then feedback will exist between them. If the concentration or effect of one variable (A—the output) is decreased in response to a change in another (B), the feedback is said to be *negative*, whereas if A increases in response to a change in B, the feedback is said to be *positive*. Hormone secretion from the posterior lobe of the pituitary (Figure 47) is controlled by fairly simple *feedback loops*.

feedback regulation*

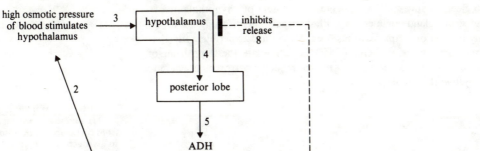

FIGURE 47 Feedback control of the secretion of antidiuretic hormone. The release of the hormone is regulated by an end-product of its own action (rather than by another hormone).

☐ Follow the feedback loop shown in Figure 47 (steps 1–8). This illustrates the order of events involved in the control of the volume of body fluid by means of ADH. Is the control of the secretion of ADH an example of positive or negative feedback?

■ It is a negative feedback loop because output (the secretion of ADH) leads to a change in osmotic pressure, which leads to a decrease in ADH secretion.

The control of the secretion of hormones from the anterior pituitary is more complex because feedback control is often exercised at both the hypothalamus and the pituitary, so two feedback loops are involved (Figure 48, *overleaf*).

These are often termed long and short loops. For example, a high level of thyroxin in the blood inhibits the release of thyroid-stimulating hormone (TSH). Inhibition takes place at two sites, the hypothalamus and anterior pituitary.

□ Look at Figure 48. Which type of feedback, positive or negative, is exemplified by the thyroxin feedback loop?

■ This is another example of a negative feedback loop.

Recent evidence indicates that different feedback loops can interact with each other. For example, high levels of corticosteroids inhibit the secretion of *adrenocorticotropin (ACTH)* by means of standard, negative feedback loops because ACTH stimulates the secretion of corticosteroids. However, high levels of corticosteroids also make cells that secrete thyroid-stimulating hormone (TSH) more sensitive to thyroid-stimulating hormone releasing factor (TRF) so that more TSH is released in response to a given quantity of TRF. Corticosteroids therefore inhibit the secretion of ACTH and stimulate the secretion of TSH.

The secretion of prolactin (PRL), growth hormone (GH) and melanocyte-stimulating hormone (MSH) is normally suppressed by release-inhibiting factors, and the control of these hormone systems seems to be more complex. For example, the secretion of prolactin release-inhibiting factor is influenced by (a) various neural stimuli and (b) the steroid hormones, progesterone and oestrogen; but there does not appear to be any simple feedback loop. To complicate matters further, thyroid-stimulating hormone releasing factor (TRF) stimulates the secretion of prolactin!

By now, you should appreciate that the pituitary and hypothalamus are key centres of physiological control. Unlike the rest of the brain, the hypothalamus is not protected by the blood–brain barrier (recall Section 4.3), so it is exposed to any physical and chemical variations that take place in the bloodstream. Various sensory devices in the hypothalamus monitor these changes. Also, the neurons of the hypothalamus are supplied with additional sensory information from various centres in the brain. In short, the hypothalamus occupies the crossroads at which neural and blood-borne information meet. It is thus uniquely placed to coordinate both sets of inputs and to activate or suppress pituitary secretions accordingly.

The arrangement, neuron → hypothalamic neurosecretory neuron → endocrine gland → endocrine gland, possessed by vertebrates, enables rapidly evoked nervous activity to be translated into sustained hormonal stimulation that results in long-term physiological changes. The sequential system of hypothalamus → pituitary → endocrine gland enables an initially weak chemical signal to be amplified many times. For example, it has been determined experimentally that 0.1 μg of corticotropin-releasing factor (CRF) can stimulate the deposition of 5 600 μg of glycogen in the liver via the adrenocorticotropic hormone–corticosteroid system (Figure 49).

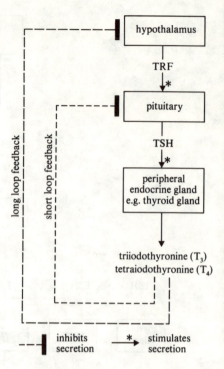

FIGURE 48 Feedback control of the secretion of tropic hormones. The thyroid hormone system is illustrated here.

FIGURE 49 Amplification in the ACTH–endocrine system. Amplification from corticotropin-releasing factor (CRF) to glycogen is $10 \times 40 \times 140 = 56\,000$ times. Blood vessels are shown in pink.

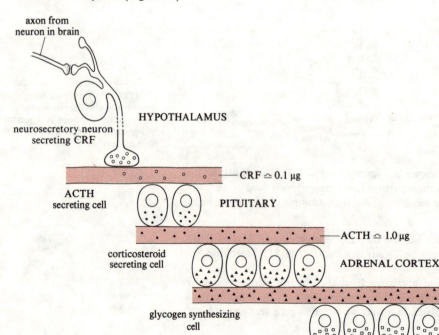

46

At each stage in the system, the secretion of a small number of signal molecules results in the release of a much larger number of signal molecules that operate at the next stage. Such amplification is necessary if the nervous and endocrine systems are to be integrated, because a few neurons, releasing minute quantities of neurotransmitter, must influence large numbers of cells (the ACTH–corticosteroid system affects virtually every cell in the body!) across what is in effect a synaptic cleft (the blood system) of enormous volume.

It is not known whether this integration of endocrine and nervous systems is widespread among the invertebrates, largely because many invertebrate endocrine systems are only just being discovered. However, insects seem to have a similar sort of system (see the TV programme, *Insect Hormones*).

7.2 Neuroendocrinology

We have seen that certain cells display both neuronal and endocrine features. You may now wonder if any endocrine glands display neuronal properties and whether hormones can influence the nervous system. Over the past few years endocrine cells of the pancreas (see Unit 17), anterior pituitary and thyroid have all been shown to produce action potentials similar to those of nerves—so the answer to the first question is yes. Thyroxin, corticosteroids and sex hormones (see Unit 18) exert extremely important influences on the brain during development. Malfunction of the thyroid gland during fetal life results in irreversible brain damage, which can lead to a condition known as 'cretinism'. The role of steroid hormones in influencing behaviour is not restricted to new-born organisms. For example, the application of minute quantities of the steroid testosterone to one area of the hypothalamus of a rat elicits maternal behaviour whereas its application to a different area stimulates male sexual activity—so the answer to the second question is also yes.

Textbooks of 20 years ago listed perhaps 20 hormones and transmitters, each with a defined location in a particular endocrine gland or brain structure. Recently, however, this division between transmitters and hormones has broken down, because several peptide messenger-molecules are now known to be secreted by both nerves and endocrine cells*. One example is growth hormone release-inhibiting factor, usually termed *somatostatin*. This is secreted by certain endocrine cells in the pancreas as well as by neurosecretory neurons in the hypothalamus and neurons in the spinal cord. Interestingly, all the peptide-producing cells involved (whether neuronal or endocrine) have a number of common metabolic features. In particular, they all take up *amines* and decarboxylate them; consequently, they are known as the *APUD* (*amine precursor uptake and decarboxylation) series of cells*. They are found scattered throughout the body, as well as in the brain, and can be identified histochemically. One hypothesis that explains these observations, and also has some experimental support, is that these cells were once all part of the same neuroectodermal region of the embryo. If this is true, the division of amino acid-based compounds into neurotransmitters and hormones is somewhat arbitrary. Also the idea of two neatly divided communication systems (nervous and endocrine) is rather misleading.

somatostatin (growth hormone release-inhibiting factor)*

amines
APUD cells

NEURON ⇆	PARANEURON ⇆ (e.g. neurosecretory cells)	PARACRINE CELL ⇆ (e.g. gastrointestinal endocrine cells)	ENDOCRINE CELL
transmit signals through combination of an action potential and a chemical signal	produce action potentials and release chemical signals that act over short distances (but longer than those of neurotransmitters)	release chemical signals that act over long distances and for considerable periods of time	
the chemical signal (neurotransmitter) acts briefly and over a very short distance	the chemical signal molecule is larger in size		
these cells possess axons	the duration of release is longer than for a conventional nerve but shorter than for conventional hormones		
	some of these cells have axons, some do not		

FIGURE 50 The neuro–endocrine continuum. Various intermediate cell types occur.

*See the Introduction to Units 16–25 for the location and function of such common peptides.

Physiologists now think in terms of a continuum (Figure 50) with conventional neurons and endocrine cells occupying opposite ends of the spectrum. In between are cells that display neuronal and endocrine features to varying extents. These may be termed *paraneurons* or *paracrine cells*.

The existence of neurons that secrete chemical messengers that act over long distances and the evidence that some endocrine cells demonstrate electrical activity similar to that of neurons suggest that it is no longer sensible to think of 'defined' nervous and endocrine systems. In the next Section, this conclusion is reinforced when we see that the effect of chemical messengers at the target—be it neuron, muscle, endocrine gland or whole organ—is similar, irrespective of the communication system that released them.

Summary of Section 7

1   Specialized neurons (neurosecretory neurons) release substances (neurohormones) into the bloodstream, which transports them to target cells.

2   Neurosecretory neurons permit the activity of the nervous system to be coordinated with that of the endocrine system. Such interactions enable the short-term activity of a few neurons to be amplified into an action of relatively long duration that can influence a large number of cells. Such systems are known to exist in both vertebrates and invertebrates.

3   The hypothalamus and pituitary are important centres of physiological regulation because within these two structures a large amount of nervous and hormonal information is integrated and used to regulate a wide range of body functions.

4   The pituitary has two parts, the anterior lobe and the posterior lobe. The latter acts merely as a releasing site (a neurohaemal organ) for neurohormones produced within the hypothalamus; the former is a complex tissue containing different types of endocrine cells. Anterior pituitary cells are controlled by neurosecretory neurons that originate in the hypothalamus and terminate in the pituitary stalk. These cells secrete neurohormones which are conveyed to the anterior pituitary by a network of blood vessels (a portal system). These neurohormones (releasing or release-inhibiting factors) either stimulate or inhibit the release of particular hormones from the anterior pituitary. Many of the anterior pituitary hormones have a tropic effect (i.e. they stimulate the release of hormones from other endocrine glands).

5   The neurosecretory cell → endocrine cell → endocrine cell arrangement exemplified by the vertebrate endocrine system produces a large amplification of the original signal. It also permits the generation of a range of feedback loops that enable the whole system to be finely regulated.

6   The hypothalamus is a unique region of the brain. It is not protected by a blood–brain barrier and is therefore subjected to all changes in the internal environment of the animal. It is connected via neurons to the central nervous system and via neurosecretory neurons to the endocrine system.

7   The endocrine and nervous systems interact: nerves control the activity of the endocrine system and the endocrine system regulates the activity of nerves.

8   The distinction between neurons and endocrine cells can no longer be regarded as clear-cut; some endocrine cells exhibit action potentials and secrete substances normally associated with neurons; similarly, some neurons secrete polypeptide hormones usually associated with endocrine cells. Consequently, neurons and endocrine cells should be regarded as opposite ends of a continuum.

Objectives and SAQs for Section 7

Now that you have completed this Section, you should be able to:

★ list four ways in which neurosecretory neurons differ from conventional neurons.

★ outline, giving specific examples, the different types of interaction that take place between the hypothalamus, pituitary and peripherally sited endocrine glands.

★ predict how levels of other hormones in the blood might change when the concentration of one hormone is altered.

★ discuss the relationship between nerves and endocrine cells.

To test your understanding of this Section, try the following SAQs.

SAQ 17 (*Objective 11*) List four ways in which neurosecretory neurons differ from conventional neurons.

SAQ 18 (*Objectives 12 and 14*) The following kinds of statement are often written hurriedly during examinations. They contain some truth but not the whole truth, and are therefore misleading. Explain briefly what the writer has omitted to say.

(a) Cells of the anterior lobe of the pituitary are connected to the hypothalamus by neurons.

(b) The secretion of hormones from the anterior lobe of the pituitary is controlled by releasing factors secreted by the brain.

(c) The secretion of corticosteroids is controlled by the pituitary.

(d) Nerves communicate via electrical impulses; in contrast, endocrine cells use hormones.

(e) There is no difference between nerves and endocrine cells.

SAQ 19 (*Objectives 12 and 13*) Apart from diabetes (see Unit 17) and infertility, malfunction of the thyroid hormone system is the commonest form of endocrine disease. Usually either an overactive (hyperthyroidism) or an underactive (hypothyroidism) thyroid gland is involved. (An overactive gland secretes too much hormone whereas an underactive gland secretes too little.) One method of rapidly distinguishing between the two malfunctions involves giving the patient an intravenous injection of synthetic thyroid-stimulating hormone releasing factor (TRF) and measuring the level of thyroid-stimulating hormone (TSH) just before and after injection. Figure 51 shows typical responses from hypothyroid, hyperthyroid, and normal individuals.

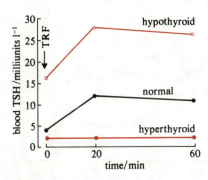

FIGURE 51 Changes in TSH levels after the injection of TRF at time zero.

From the Figure and what you have learned about hormone feedback systems, answer the following.

(a) Suggest why the level of TSH in the blood of hyperthyroid patients does not rise.

(b) Why is the pre-injection level of TSH in hypothyroid patients so high?

(c) What would you infer about the possible sites of malfunction if an individual appeared normal on the TRF test but had previously been diagnosed as hypothyroid on the basis of very low levels of thyroxin (T_4) and triiodothyronine (T_3) in the blood?

(d) In the patient described in (c), there are two possible sites where a malfunction may be occurring. Decide where these are and then suggest which hormones should be measured to determine the precise site of malfunction.

SAQ 20 (*Objective 13*) Organ transplant patients are often given massive quantities of corticosteroids. These are used because they suppress the body's natural defence mechanism (its immune system) and so prevent the rejection of the foreign tissue. From the information in this Unit, predict how this therapy will affect the blood levels of at least three other hormones.

8 The action of hormones and neurotransmitters at the cellular level

This Section, though quite short, is important. The most difficult ideas conceptually are the regulation of receptors and the actions of second messengers (summarized in Figure 55).

Similarities between nervous and endocrine control are very marked at the biochemical level. All the 'language' used in communication (hormones or neurotransmitters) is ultimately chemical in nature and, consequently, the molecular mechanisms involved in the receipt and translation of hormonal and neuronal messages by target cells are often indistinguishable. The detection and translation of these chemical messages usually involve three basic steps (Figure 52).

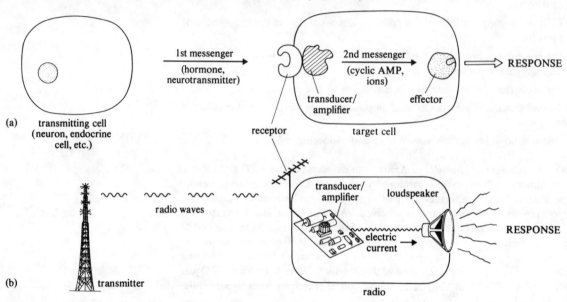

FIGURE 52 An intercellular communication system (a) compared with a more familiar communication system (b). Note how similar the two are in terms of their basic components.

1 An initial *binding* of the intercellular messenger molecule (usually termed the *first messenger*) in a highly specific fashion to *receptor molecules* situated in the membrane or cytoplasm of the target cell.

first messenger*
receptors*

2 A *transduction* step in which the interaction of the first messenger and receptor triggers the 'production' of a second set of *intracellular*, chemical, signalling substances (termed *second messengers*). *Amplification* of the initial signal usually occurs at this stage because the interaction of *one* first messenger molecule with an appropriate receptor can produce *many* molecules of a second messenger.

second messenger*

3 A final *effector* step in which the second messengers act directly (or indirectly) to activate (or inactivate) a specific, cellular, effector system that produces the observed cellular response.

8.1 Receptors

A *receptor molecule* can be thought of as a molecule 'aerial' that is 'tuned' to listen on a particular molecular 'waveband'. In chemical terms, each type of receptor has a unique, three-dimensional structure that incorporates a highly specific binding site; this ensures that the cell responds only to those first messengers for which it possesses receptors.

☐ What class of biological molecule is most likely to be used in the construction of receptor molecules?

■ You should have been able to answer this straight away from what you have already learned about enzymes and transport in Units 6, 7 and 9 and 10. The answer is, of course, proteins of various kinds. Not surprisingly, most receptor molecules possess many of the properties exhibited by enzymes and transport molecules and can be characterized by K_d, a measure of the affinity of binding, and v_{max} (B_{max}), a measure of the density of binding sites.

K_d, measure of affinity of binding*
v_{max}(B_{max}), density of binding sites*

TABLE 4 The characteristics of receptor molecules for three different messengers

Receptor for	Chemical nature	Relative molecular mass	No. per cell	K_d value/mol l^{-1}
acetylcholine	glycolipoprotein arranged in several subunits	88 000–500 000	depends on the cell, but can reach values as high as 10^{11}	$\simeq 2 \times 10^{-8}$
thyroid-stimulating hormone (TSH)	glycoprotein arranged in several subunits	280 000	500 (thyroid cells)	1.9×10^{-9}
insulin	glycoprotein arranged in several subunits	300 000	75 000 (liver cells)	1.2×10^{-6}

Table 4 contains some detailed information on three receptors; *do not attempt to memorize the actual details in the Table*. As with enzymes and transport molecules, the rate of interaction of first messenger and receptor molecules depends largely on the chances of one type of molecule meeting the other. During stimulation of the endocrine or nervous system, the concentration of first messengers rises in the immediate vicinity of the cell; consequently, there will be an increased chance that a first messenger molecule will interact with a receptor molecule.

As the concentration of first messenger molecules rises, more and more receptor binding sites are occupied, until eventually all the receptor sites are full. Generally, but not invariably, the observed cellular response is proportional to the number of receptor sites occupied.

So far, we have tacitly assumed that cells possess fixed numbers of particular receptor molecules. What would happen if a cell could adjust the number of its receptor molecules? To answer this question, we need to know what would happen to the number of receptors at any instant in time, if the extracellular concentration of first messenger molecules was fixed and the density of receptor molecules (i.e. the number per unit area) was suddenly increased *or* decreased.

☐ In normal humans the concentration of triiodothyronine (T_3) in the blood remains fairly constant at about 1.7 ng cm^{-3}, and at this concentration, 40–50 per cent of T_3 receptors in liver cells are occupied with T_3. Predict in general terms how (a) *increasing* and (b) *decreasing* the density of receptor molecules would affect the reaction of a liver cell to T_3.

■ Increasing the density of receptor molecules while holding the concentration of first messenger steady ought to result in the occupation of a greater number of receptor sites because this increases the chances of a hormone molecule (or neurotransmitter) meeting a receptor. Decreasing the density of receptors should have the opposite effect. On this basis, the net effect of a target cell's increasing its receptor population would be to increase the sensitivity of the cell to a particular hormone, whereas decreasing the density of receptors would decrease the cell's sensitivity.

Target cells that are exposed to high levels of hormones or neurotransmitters for long periods of time tend to become *less* sensitive to first messenger molecules, and higher concentrations are needed to produce a response. Conversely, target cells that have not been stimulated by appropriate first messengers for some time tend to become *more* sensitive, and lower concentrations of first messenger elicit a maximal response (e.g. insulin; see Unit 17). Such an alteration in sensitivity can often be explained on the basis of changes in the number of receptors, but this is not invariably the case.

Exactly how such changes in *sensitivity* are controlled is only now beginning to be understood. In some cases sensitivity to a particular first messenger may be modified by other messenger molecules. For example, oestrogens make the chick oviduct (Units 12 and 13, Section 8.2) more sensitive to progesterone by stimulating the synthesis of progesterone receptors.

regulation of receptors (sensitivity)*

The ability of target cells to change their sensitivity is of physiological importance because it provides a fail-safe component in the communication system that can compensate to some extent for malfunctions of various sorts. In certain feedback loops, the sensitivity of target cells can be altered continually as a result of the interplay between various control systems. Target cells have different degrees of sensitivity at different stages of their lives.

If another molecule with a very similar structure is present in appreciable concentrations it can occasionally interfere by replacing the normal molecule and triggering the same response. Such a molecule is termed an *agonist*. Alternatively the interfering molecule may simply occupy the receptor preventing it from combining with another normal or agonist molecule and thus blocking any response. In this case the molecule is referred to as an *antagonist*. Agonists and antagonists may be substances that occur naturally, or they may be drugs that are specifically manufactured for their special triggering or blocking properties (Figure 53).

agonist*

antagonist*

AGONIST

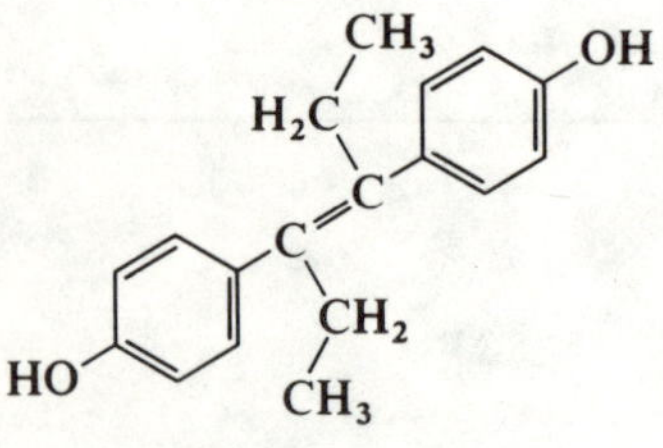

diethylstilboestrol

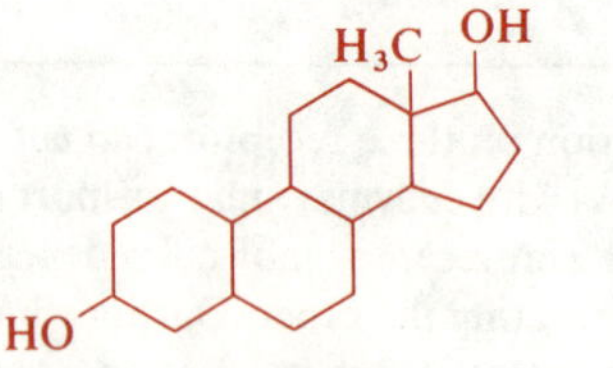

oestradiol

AGONIST

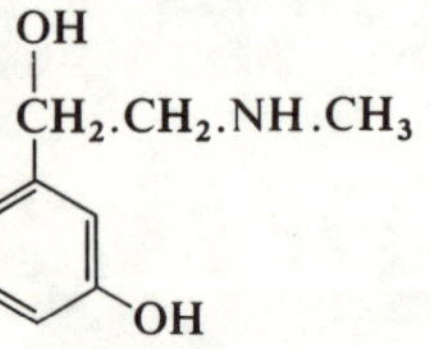

phenylephrine

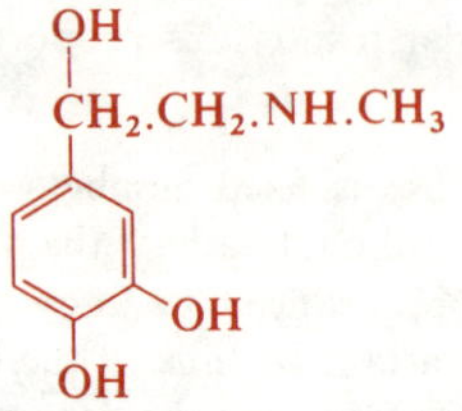

adrenalin

ANTAGONIST

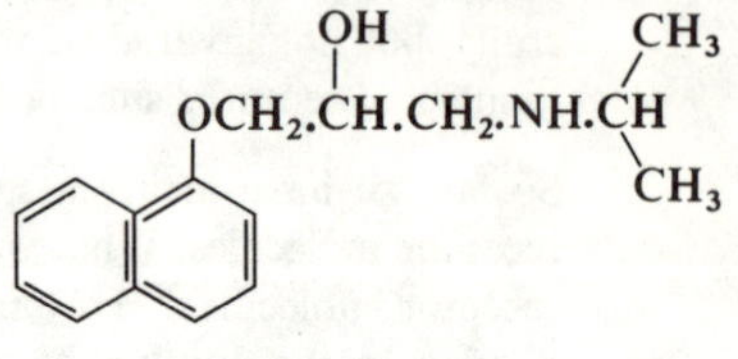

propranolol

FIGURE 53 Some examples of synthetic agonists and antagonists. (Natural hormones are shown in red for comparison.) Phenylephrine (an agonist) mimics adrenalin and is used in various 'cold-relief' treatments because of its ability to prevent excessive secretion of mucous. Diethylstilboestrol mimics many of the actions of oestradiol (and is cheaper to make!). It has been used as an additive in cattle food to promote weight gain. Propranolol blocks the response of the heart to adrenalin and noradrenalin.

□ We mentioned an agonist effect earlier in this Unit. Which two hormones are able to interact with each other's receptors and why do they usually not interfere with each other?

■ Oxytocin and antidiuretic hormone (ADH). They do not usually interfere with each other because the ADH receptor has a higher affinity for ADH than for oxytocin.

So far, we have avoided discussing the location of particular receptors. However, you ought to be able to work this out from what you know about transport across membranes.

□ Where in cells would you expect to find receptors for (a) protein hormones/neurotransmitters and (b) steroid hormones? (c) By what technique would you test your hypothesis?

■ (a) Proteins and amino acid derivatives (neurotransmitters) are hydrophilic molecules. They would not cross the cell membrane easily, so the likely site for such receptors is on the outside of the membrane.

(b) Steroids are derived from a lipid molecule (cholesterol) so you should expect steroids to cross membranes readily (recall Unit 5, Section 8). The receptors could therefore be sited either on the outside of the membrane or in the cell cytoplasm. The majority of steroid receptors are cytoplasmic but some recent evidence suggests that some cells may possess membrane-bound steroid receptors.

(c) The location of particular receptors can be demonstrated by the technique of autoradiography using radioactively labelled hormones or neurotransmitters (see Unit 4, Section 4.1).

8.2 Second messengers

The interaction of first messenger and receptor is translated into intracellular action via the agency of one (or more) of three different types of second messenger substances: messenger RNAs, *cyclic nucleotides*, and inorganic ions.

cyclic nucleotides (e.g. cAMP)*

8.2.1 Steroid and thyroid hormones

Current evidence suggests that cytoplasmic steroid receptors are probably composed of two subunits, each of which binds one steroid molecule (Figure 54).

When both receptor sites R_A and R_B are occupied by the appropriate steroid, the hormone–receptor complex migrates to the cell nucleus. The ensuing events are most completely worked out for the progesterone receptor found in chick oviduct cells, but other steroid hormones appear to have a similar mode of action. The two subunits of the progesterone receptor are not identical and have different functions, although they are similar in size. One subunit (subunit B) ensures that the hormone–receptor complex binds to a specific acidic protein that forms

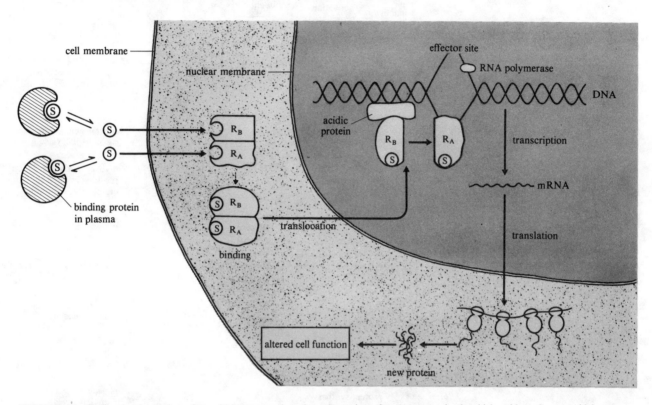

FIGURE 54 The molecular action of steroid hormones, based on the action of progesterone in the chick oviduct. S = steroid hormone. R_A and R_B are the receptor subunits: R_A is responsible for activation and R_B is responsible for binding.

part of the chromatin. Once the dimer has bound to the correct region of the chromatin, A and B dissociate and the A subunit then interacts directly with the adjacent region of the DNA molecule. This results in the *transcription* of this section of DNA and the subsequent production of many mRNA molecules. In this system, mRNA molecules constitute the second messengers, and polysomes are the effectors responsible for the cellular response. The thyroid hormones seem to act in a similar manner except that the receptor is sited permanently in the nucleus.

transcription

8.2.2 Polypeptide hormones and neurotransmitters

In contrast to steroid and thyroid hormones, most protein hormones and neuro-transmitters seem to exert their effects by quite different mechanisms. In most cases the interaction of either of these substances with appropriate receptors in the cell membrane results in either (i) the activation of membrane-bound enzymes or (ii) the opening of simple ion gates in the membrane.

In the first system the activated enzyme is normally *adenylate cyclase* which catalyses the conversion of ATP to cyclic adenosine monophosphate (cyclic AMP or cAMP) (Figure 55a). A single activated adenylate cyclase molecule can produce about 1 000 cyclic AMP molecules per minute, so the enzyme amplifies the initial interaction of first messenger and receptor and converts this event into an intracellular signal. In this sort of system, cyclic AMP acts as the second messenger and exerts its effects primarily by activating a set of enzymes known as *protein kinases*, which in turn phosphorylate proteins (Figure 55a). These proteins can be

adenylate cyclase*

protein kinases*

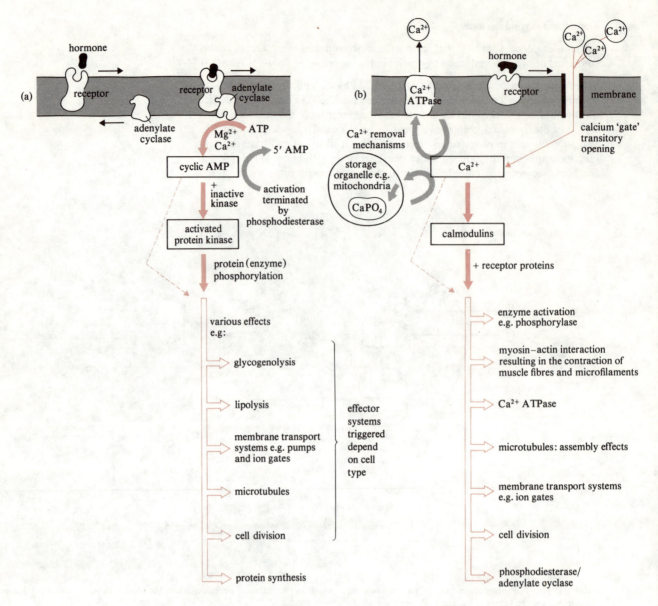

other enzymes, structural proteins (e.g. ribosomes and membrane proteins) or nuclear proteins such as histones (recall Unit 8, Section 3, and see Units 17 and 18). Phosphorylation of a protein may result in it adopting a different steric configuration. For enzymes this change in shape may underlie the activation or inhibition of catalytic properties; for membranes it may result in changes in permeability to ions (opening or closing of simple ion gates); for ribosomes it may alter their translational properties; and for nuclear proteins it may affect the expression of DNA.

In the second system (Figure 55b), the interaction of first messenger and receptor usually results in the opening of Ca^{2+} channels in the membrane.

☐ The intracellular level of Ca^{2+} is normally between 10^{-5} and 10^{-7} mol l^{-1}, whereas the concentration of Ca^{2+} in the medium bathing most cells is usually about 10^{-3} mol l^{-1}. What will happen when Ca^{2+} ion gates are opened in response to stimulation by first-messenger molecules?

■ Ca^{2+} will flood into the cell down its concentration gradient, provided there is no significant opposing electrical gradient. Not only will this cause a rise in intracellular calcium but it may also cause a change in the difference in potential across the cell membrane (Units 9 and 10, Section 2).

Because the opening of a simple ion channel results in an influx of Ca^{2+}, the original interaction of first messenger and receptor is amplified and Ca^{2+} serves as the second messenger or intracellular signal. Many cells are known to possess *calmodulins* (regulatory proteins that are dependent on calcium), which are analogous to the protein kinases that are dependent on cyclic AMP. Both sets of

FIGURE 55 (a) The cyclic AMP system. A messenger molecule causes a change in the conformation of the receptor that enables it to interact with, and activate, adenylate cyclase. The receptor and cyclase react by virtue of 'fluid' movement within the membrane. The cyclic AMP produced initiates a wide range of cellular changes, usually via protein kinases (a few *direct* effects are known).

(b) The calcium–calmodulin system. Note that storage organelles and calcium ATPase perform a similar function to phosphodiesterase in the cyclic AMP system. Exactly which of the effects shown are direct actions of Ca^{2+} and which require the agency of calmodulins is not known as yet.

calmodulins*

proteins are controlled by second messengers and are able to produce a wide range of effects (Figure 55), depending on the way the particular target cell is programmed to respond. In addition, Ca^{2+} can act directly at allosteric sites on particular key enzymes, and/or the concomitant changes in membrane potential that accompany an influx of ions can be used to trigger voltage-dependent ion gates, which will result in the influx of other ions (e.g. Na^+ and K^+).

Just as there are elaborate, physiological feedback mechanisms controlling the level of blood-borne hormones, so mechanisms exist for regulating the intracellular levels of second messengers.

In the cyclic AMP system (Figure 55a), in unstimulated cells adenylate cyclase molecules are, so to speak, just ticking over and a small amount of cyclic AMP is produced. However, under these conditions, the cellular level of cyclic AMP never rises because cells also possess a set of enzymes (located in the cytoplasm) known as *phosphodiesterases* that degrade cyclic AMP to inactive AMP. Consequently, cyclic AMP is degraded as fast as it is produced. When a cell is stimulated, the rate of production of cyclic AMP increases dramatically (e.g. in the turkey, the production of cyclic AMP is increased ten-fold when adenylate cyclase molecules in the membranes of red blood cells are stimulated). Consequently, the capacity of the phosphodiesterase system is soon exceeded and the intracellular level of cyclic AMP begins to rise, triggering the cell's response. Once the adenylate cyclase molecules are inactivated, the level of cyclic AMP falls because the phosphodiesterase is once more able to remove cyclic AMP molecules faster than they are produced.

phosphodiesterases

The calcium system (Figure 55b) operates by quite a different mechanism, although the basic principle is the same. Calcium pumps in membrane-bound organelles (particularly mitochondria) and in the cell membrane (e.g. Ca^{2+} ATPase) are used to reduce the level of Ca^{2+} in the cytoplasm.

□ The K_d value for Ca^{2+} uptake in mitochondria is 100 mmol l^{-1}. What does this indicate about the transport system? (Most ion transport systems have a K_d of 2–30 mmol l^{-1}.)

■ The K_d value is relatively high, so the affinity of the uptake system is low; therefore its efficiency is low. This system will therefore operate in a similar fashion to phosphodiesterase.

The remarkable feature of these two intracellular, second-messenger systems is that a large proportion of all cellular responses induced by polypeptide hormones and neurotransmitters can be explained on the basis of changes in intracellular levels of either Ca^{2+}, or cyclic AMP, or both (Table 5). In essence, one type of target cell differs from another type not in terms of its second-messenger system(s) but in terms of the receptors that are capable of activating the second-messenger system. Consequently, the action of many hormones/neurotransmitters can be mimicked by either artificially raising the level of cyclic AMP and/or causing an influx of calcium with various drugs.

TABLE 5 Responses of some tissues to electrical and chemical stimuli

Tissue	Stimulus	Response	Ca^{2+} required	cAMP produced
synapses	depolarization	release of transmitter	+	+
anterior pituitary	GRF	release of growth hormone	+	+
anterior pituitary	FSH/LH-RF	release of LH	+	+
anterior pituitary	TRF	release of TSH	+	+
posterior pituitary	depolarization	release of ADH	+	?
B cell, pancreas	glucose	release of insulin	+	+
adrenal cortex	ACTH	release of corticosteroids	+	+
adrenal medulla	depolarization	release of adrenalin	+	?
liver	glucagon	synthesis and release of glucose	?	+
thyroid	TSH	release of thyroxin	+	+
corpus luteum	LH	release of progesterone	+	+
heart	adrenalin	glycogenolysis	+	+
adipose cell	adrenalin	lipolysis	+	+

In some target cells the Ca^{2+} and cyclic AMP second-messenger systems act antagonistically, one second messenger switching on an effector system in response to a particular first messenger, while the other second-messenger system switches off the effector in response to a different hormone or neurotransmitter. For example, Ca^{2+} causes a smooth muscle cell to contract whereas cyclic AMP causes it to relax. In other situations the two second-messenger systems act in a *synergistic* (coordinated) fashion to produce an overall response (Figure 56). For example, the release of adrenalin from the adrenal medulla as a result of stimulation by acetylcholine is mediated by Ca^{2+} (A). Medulla cells *also* have receptors linked to adenylate cyclase (B), and a rise in the level of cyclic AMP (C) causes a release of Ca^{2+} from stores within the cells (D). In the adrenal medulla, then, cyclic AMP acts synergistically with the Ca^{2+} influx produced by acetylcholine. If the level of cyclic AMP rises higher, the synthesis of adrenalin is stimulated (E) as well as the release of calcium. The receptor concerned in these responses is activated by adrenalin! This produces a rather neat system for ensuring that the release of adrenalin is rapid and massive and, in addition, that depleted stores of adrenalin are then replenished. This synergistic behaviour of messenger molecules is something we shall return to in Units 17 and 18.

synergism*

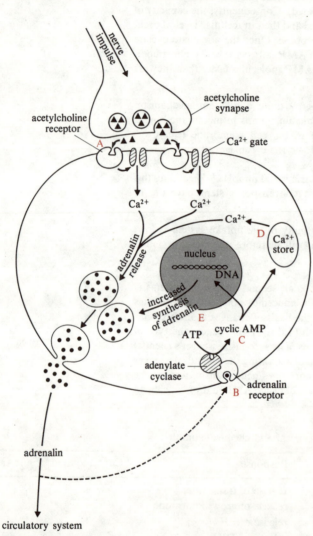

FIGURE 56 Various factors that affect the release and synthesis of adrenalin from the adrenal medulla cell.

Complex feedback interrelationships are known to exist between the cyclic AMP and Ca^{2+} second-messenger systems (e.g. Ca^{2+} can alter the activity of adenylate cyclases and phosphodiesterases), but these are beyond the scope of this Course. Other second messengers are also implicated in some responses, but usually have a secondary role. At nerve synapses, sodium and potassium ions can be thought of as second messengers, controlling voltage-dependent ion gates by changing the E_m of the nerve cell membrane (recall Section 4).

Summary of Section 8

1   Changes in the activity of target cells in response to hormonal or neuronal signals involve the specific binding of a hormone or neurotransmitter (first messenger) to a receptor molecule sited on the membrane or within the cell. This results in the production of a second-messenger molecule that interacts with cellular effectors (e.g. enzymes, microfilaments, etc.), the overall response depending on the particular type of effector(s) possessed by the cell.

2   The sensitivity of a cell to a particular hormone or neurotransmitter can be modified by changes in the density of receptor molecules. Such changes are regulated by a variety of factors (e.g. hormone levels in the blood, the developmental stage of the cell, etc.). A decrease in the density of receptors effectively decreases the sensitivity of the cell to a particular hormone or neurotransmitter.

3   The action of a particular hormone or neurotransmitter can often be (a) blocked or (b) mimicked by structurally similar chemical compounds; these are (a) antagonists and (b) agonists.

4   Receptors for neurotransmitters and polypeptide hormones are usually situated on the outside of the cell membrane, whereas receptors for steroid and thyroid hormones are usually sited within the cell cytoplasm or nucleus.

5   Binding of hormones (or neurotransmitters) to receptors is characterized in the same way as the interactions of enzymes and their substrates and ions and their transport molecules, that is, using v_{max} (B_{max}) and K_d.

6   The transduction/translation of the initial stimulus into a second messenger has two functions: (a) it converts the extracellular signal into an intelligible intracellular signal; (b) it serves to amplify the extracellular signal.

7   Cyclic AMP and Ca^{2+} are the most commonly used second messengers. These may act in conjunction with each other to produce an overall effect or they may act in opposition, one switching on a particular response and the other switching it off.

8   Concentrations of second-messenger molecules are regulated by various factors in a way similar to the regulation of levels of first-messenger molecules. The most significant regulators are: phosphodiesterase activity for cyclic nucleotides, and ion pumps for inorganic ions.

9   Because many first messenger systems use the same second messengers, it is possible to mimic the cellular actions of many hormones or neurotransmitters by artificially raising the levels of cyclic nucleotides and/or ions within the cell.

Objectives and SAQs for Section 8

Now that you have completed this Section, you should be able to:

★   list the major steps between the release of a messenger molecule and the response elicited in a target cell.

★   interpret simple data relating to the kinetics of binding messengers to target tissue.

★   describe with simple diagrams the mechanisms through which *two* second messengers operate.

★   design a simple experiment to test whether a drug is an agonist or an antagonist of a chemical messenger.

To test your understanding of this Section, try the following SAQs.

SAQ 21   (*Objective 15*)   The binding of hormones and receptors can be characterized in more or less the same way as interactions of enzymes and substrates or transport molecules and substrates. The K_d for the binding of T_3 (triiodothyronine) to thyroid hormone receptors in the nucleus is 1×10^{-11} mol l^{-1}, whereas for T_4 (thyroxin) the K_d is 1×10^{-10} mol l^{-1}. Explain what this difference indicates and suggest which of the two hormones would be the more effective.

SAQ 22 (*Objective 15*) Figure 57 shows the results of some experiments on isolated pituitary cells that secrete thyroid-stimulating hormone (TSH). In these experiments, radioactive thyroid-stimulating hormone releasing factor (TRF) was used to assess the number of TRF receptors in a standard quantity of tissue. The technique is fairly simple: cells are incubated for a set period with a known quantity of radioactively labelled TRF (during this period the radioactive TRF binds to any TRF receptors in the cells); the cells are then washed vigorously several times under carefully controlled conditions to remove any unbound TRF.

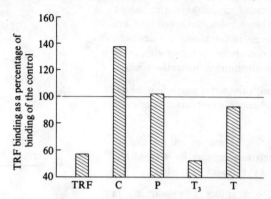

FIGURE 57 The effect on subsequent TRF binding of previously incubating cells with various hormones (control cells were incubated in plain saline solution). Binding is expressed in relation to binding of the control cell ($=100\%$).

The number of receptors possessed by a specific quantity of tissue can then be assessed by measuring the radioactivity of the washed tissue. A highly radioactive tissue contains many receptors and a very weakly radioactive tissue sample only a few. In these experiments the quantity of TRF bound by (a) a set of control cells and (b) sets of cells that have been incubated for 42 hours with various hormones. The control cells were not previously incubated with any hormone.

What can be deduced about the effect of treatment with TRF, cortisol (C), testosterone (T), triiodothyronine (T_3), and progesterone (P) on subsequent binding of TRF, and hence about the number of receptors and the cells' sensitivity to TRF?

SAQ 23 (*Objectives 15 and 16*) Z is a natural messenger molecule that causes the manufacture of a particular protein in rat liver cells through the agency of cyclic AMP. You wish to produce, synthetically, a substance very like Z, with the same effect, and you end up with five compounds (A–E) that are similar to Z in structure. How would you set about testing your five candidates?

SAQ 24 (*Objectives 15 and 16*) Carrying out the experiments in SAQ 23 gives the results in Table 6. Classify each of the five compounds (A–E) as an agonist or an antagonist or neither.

TABLE 6 The effects of substances Z and A–E on the level of cyclic AMP and on protein production in liver cells

Substance	Production of liver protein (% increase)	Change in cell cAMP levels (% increase)
Z alone	100	100
A alone	no change	no change
B alone	50	50
C alone	10	no change
D alone	no change	10
E alone	100	100
Z plus A	no change	no change
Z plus B	110	110
Z plus C	100	100
Z plus D	100	100
Z plus E	150	150

9 Concluding remarks

You should now see clearly the basic principles of physiological communication. A stimulus from within or outside the organism triggers the release of a chemical messenger that communicates with specific target cells through receptors for that chemical messenger. These receptors are found in or on the cell and only those cells with appropriate receptors will respond to a particular message. The nature of the target cell's response depends on what function that cell is specialized to perform (e.g. secretion, or contraction, or generation of nerve impulses). The way the response is effected is through a second-messenger (amplification) system which is linked to the receptor. The major second-messenger systems appear to be cyclic nucleotides and calcium ions, which can interact closely with one another in a target cell. In other cells (e.g. some neurons and their dendrites) the second-messenger system uses monovalent ions such as Na^+, K^+ and Cl^-.

Throughout the Unit we have emphasized the similarities between the two major communication systems (nerves and hormones), not only in their use of chemical messengers (many of which are common to both systems) but also in the way that messengers and receptors interact. Between them, nerves and hormones can achieve fine regulation, and control short-term and long-term events. Figure 58

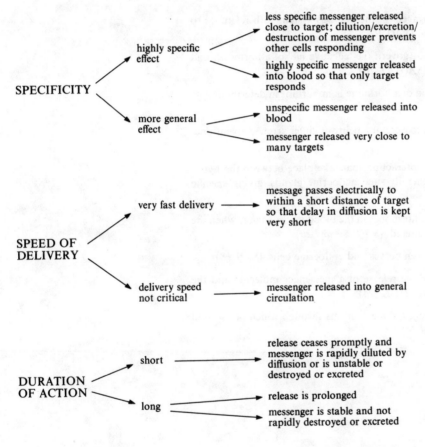

FIGURE 58   A summary of the different ways in which communication systems act to produce a variety of results.

attempts to summarize the ways in which communication systems do this, and you can see that nowhere are the words 'nerve' or 'hormone' used specifically. The overlap in the involvement of these two systems makes such a precise allocation of roles unproductive.

Unit 16 introduced the basic principles. Units 17 and 18 are concerned with two important and rather different physiological processes: the regulation of blood glucose and the control of reproduction. These processes illustrate the many ways in which communication systems interact to produce precise effects, often over considerable periods of time.

Objectives for Unit 16

1 Define and use, or recognize definitions and applications of the terms marked by an asterisk in Table A.

2 Identify the basic components of an intercellular communication system. (*SAQ 1*)

3 Describe the basis of the membrane potential, local potential, action potential, and synaptic potential in terms of changes in permeability to ions. (*SAQs 2, 4 and 5*)

4 Describe the relationship between fibre size and the velocity of conduction in nerves. (*SAQ 3*)

5 Describe the major events between the arrival of one action potential in an axonal ending and the generation of another in the axon hillock of a receiving neuron. (*SAQs 6 and 7*)

6 List at least three examples of the functional relationship between neurons and glial cells. (*SAQ 8*)

7 Describe ways in which information can be coded in the nervous system. (*SAQ 9*)

8 Show how circuits can be 'engaged' in the nervous system with reference to simple and more complex reflexes. (*SAQ 10*)

9 Outline the various factors that govern the 'effective life' of a hormone, and hence the duration of its action. (*SAQs 11–16*)

10 Explain why the concentration of a hormone is important in determining its target tissue. (*SAQs 11–16*)

11 List four ways in which neurosecretory neurons differ from conventional neurons. (*SAQ 17*)

12 Outline the different types of interaction that take place between the hypothalamus, the pituitary and peripherally sited endocrine glands, giving specific examples. (*SAQs 18 and 19*)

13 Predict how the levels of other hormones in the blood might change when the concentration of one hormone is altered. (*SAQs 19 and 20*)

14 Discuss the relationship between nerves and endocrine cells. (*SAQ 18*)

15 List the major steps between the release of a messenger molecule and the response elicited in a target cell. (*SAQs 21–24*)

16 Describe, with simple diagrams, the mechanisms through which *two* second messengers operate. (*SAQs 23 and 24*)

SAQ answers and comments

SAQ 1 T, X and N are neuron cell bodies; S and W and M are dendritic (receptive) zones; V and Z are synapses; U and Y are axons. The three correct items are therefore (iv), (v) and (vii). If you chose any other items, quickly re-read Section 2.2.

SAQ 2 Action potentials would be generated during the first part of the experiment (ii). The generation of action potentials is a passive process and therefore does not require ATP. After some time, the metabolic inhibitor would limit the amount of ATP available in the axon to restore the levels of Na^+ and K^+ via the Na^+–K^+ exchange pump. As a result action potentials would be progressively harder to generate as Na^+ accumulated in the axon and the E_m approached zero.

SAQ 3 (a) Species X is probably an invertebrate because the axon has a small diameter and a low velocity of conduction. The axon of species Y is a giant fibre (500 μm in diameter) and is therefore from an invertebrate. Species Z is a vertebrate; it has axons of small diameter and high velocity of conduction.

(b) Species Z has an axon of small diameter but high velocity of conduction, which indicates that it is a myelinated axon, and therefore insulated.

(c) Increasing the size of fibres in unmyelinated axons decreases their resistance to the flow of current; therefore the ratio of membrane resistance to axon resistance increases. This results in a higher velocity of conduction because ions are less likely to leak out of the membrane.

SAQ 4 (a) An increase in permeability to Na^+ would *depolarize* the membrane and E_m would approach E_{Na^+}, which is $+45\,mV$.

(b) An increase in permeability to Cl^- would tend to hyperpolarize the membrane because E_{Cl^-} is lower than E_m.

(c) After the membrane was ruptured, the selectively permeable barrier that produced the E_m would be destroyed.

(d) A rise in the concentration of external K^+ would *depolarize* the membrane because the value of E_m is dependent on E_{K^+}. In turn E_{K^+} depends upon the difference in the concentration of K^+ outside and inside the axon.

SAQ 5 The main difference between a local potential and an action potential is in the activation of voltage-dependent ion gates for Na^+. A local potential does not reach the E_m threshold at which Na^+ gates start to open, and so it simply decays with distance. An action potential triggers Na^+ gates and this drives E_m towards E_{Na^+}. This change in the E_m is sufficient to trigger a similar overshoot all the way along the axonal membrane.

SAQ 6 Tetrodotoxin (TTX) blocks the Na^+ voltage-dependent gate. (i) The release of a transmitter at nerve–muscle junctions opens channels for Na^+, but these are insensitive to TTX. A postsynaptic potential would therefore still be recorded when the axon is stimulated. (ii) TTX would block the voltage-dependent Na^+ gates in the axon and inhibit the generation of action potentials. No change in postsynaptic potential would be recorded.

SAQ 7 Synapse A will result in a deflection of the E_m towards zero. Synapse B will result in a larger deflection of the E_m towards zero (depolarization) because the membrane's permeability to Na^+ alone is changed; that is, E_m will move towards E_{Na^+}.

SAQ 8 This question relates to Section 4.3. The correct statements are (iv) and (v). The blood–brain barrier is selectively permeable, but not only to gases: substances such as amino acids and sugars can penetrate the brain along with certain drugs. No glial cells receive structural inputs from neurons, but all of them have an E_m. Glial cells not only act as regulators of the environment of the nervous system; they also form insulating layers and appear to synthesize macromolecules for use by neurons.

SAQ 9 Sensory inputs can be coded in a number of ways: (i) according to the frequency of impulses (which are proportional to the strength of the stimulation); (ii) according to the duration of the firing of impulses (which is proportional to the duration of the stimulation); (iii) according to changes from a base firing frequency and (iv) according to the way the sensory neuron is connected with other parts of the nervous system.

SAQ 10 Your flow diagram should look something like Figure 59. When you prick your finger, three distinct actions often result:

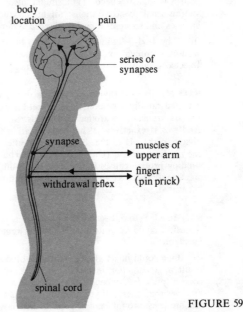

FIGURE 59

(i) the *position* of the stimulus on the body is recognized (i.e. information travels to the brain);

(ii) the finger (and sometimes the entire arm) is withdrawn from the stimulus object (i.e. information travels to the spinal cord to effect muscle movement; often the movement involves more than one set of muscles);

(iii) a sensation of pain is experienced (i.e. the brain (and often the vocal cords!) is involved).

SAQ 11 (a) True, because they are both synthesized from amino acids.

(b) True, because cortisol and progesterone are both C_{21} steroids and therefore have very similar chemical structures (see Figure 30). Growth hormone, being composed of 190 amino acids has a quite different structure.

(c) False. They also bind to binding proteins in the plasma (see Section 6.1.4).

(d) False. LDH occurs in several different forms (recall isoenzymes; Unit 6, Section 6). Particular peptide hormones (e.g. oxytocin) show a similar *polymorphism* (many forms), so it is likely that growth hormone differs slightly in different species. The function of a protein is not necessarily determined by the entire amino acid sequence—only a small section of the molecule may be involved.

(e) True; see Section 6.1.2 and Figure 33.

(f) True; see Section 6.1.2.

SAQ 12 The four factors are: (i) the presence of suitable receptors at target cells; (ii) the rate at which the hormone is secreted; (iii) the rate at which the hormone is broken down in the liver and at the target tissue; (iv) the rate at which the hormone is excreted via the kidney. Factors (ii)–(iv) control the concentration of the hormone. (See Figure 36 and Sections 6.1.4 and 6.1.5.)

SAQ 13 (a) Hormone secretion may vary over a 24 hour period (see Figure 34), so a 'normal hormone level' at, say 9 a.m., may be quite different from the normal level at 11 p.m.

(b) Small hormones, for example steroids and very small peptides, are likely to appear in the urine.

(c) Changes in the amount of hormone excreted indicate changes in the rate at which hormones are secreted or changes in the amount of protein available in the plasma for binding to the hormone. You might also have suggested that the urine could be analysed for the products of hormone metabolism. This can also yield information about changes in hormone secretion or the rate at which they are broken down.

SAQ 14 (a) If the concentration of insulin is high, it is likely to cross-react with other hormones producing potentially harmful pharmacological effects. (You might also have suggested that supranormal concentrations might produce supranormal responses and so upset a finely balanced system.)

(b) The body's immune system might classify the non-human insulin as foreign and produce antibodies that combine with the injected insulin and reduce its effectiveness.

SAQ 15 Because thyroxin is very small one would expect it to be excreted rapidly. However, its long half-life suggests that instead it may be transported around the bloodstream bound to a protein. And this is exactly what happens. Heating the blood sample or changing the pH of the solution would alter the conformation of the binding protein (Unit 5, Section 3). This denaturation of the binding protein causes it to release the bound thyroxin. This pre-treatment is necessary if the total concentration of hormone is to be measured.

SAQ 16 (a) No, because the pancreas is a complex tissue and this effect could be quite unrelated to the removal of the source of glucagon.

(b) One could inject glucagon into the bloodstream. This should result in a *reduction* in blood glucose levels if the deduction is correct (provided, of course, that the pancreas is not the organ that removes the sugar!). If you actually did this experiment the blood glucose levels would *increase* even further showing that the deduction was wrong. (Another hormone (insulin) is responsible for depressing the level of glucose in the blood. Insulin is also secreted by the pancreas, hence the elevated level of glucose after its removal.)

SAQ 17 (i) They possess more nerve endings per nerve. (ii) Their action potentials develop more slowly and are of longer duration. (iii) Their cytoplasm contains electron-dense granules that are much larger than those seen in conventional neurons. (iv) Their products are released into the blood system rather than into a conventional synaptic cleft (see Section 7).

SAQ 18 (a) The writer has omitted to mention that the connection is *indirect* (via a portal blood system) and involves *specialized* (neurosecretory) neurons that secrete neurohormones (see Section 7.1).

(b) Release-inhibiting factors are also involved in some cases. Both types of neurohormone are released from a *distinct region of the hypothalamus*, not the whole brain (see Section 7.1).

(c) The *hypothalamus* is also involved (see Section 7.1 and Figure 48).

(d) Communication via nerves usually also involves the secretion of 'hormone-like' substances (*neurotransmitters*), which in some cases are identical to hormones secreted by certain endocrine cells. In addition, *some endocrine cells can produce electrical impulses* in a fashion similar to nerves (see Section 7.2).

(e) This over-emphasizes the case set out in this Unit. It is true that some neurons are so similar to endocrine cells (and vice versa) that it is difficult to decide what to call them. However, some neurons are easily distinguished from endocrine cells. It is better to talk in terms of neurons and endocrine cells being very closely related and at opposite ends of a continuum of cells (see Section 7.2).

SAQ 19 (a) These individuals will have high levels of thyroxin (T_4) and triiodothyronine (T_3) in their blood. Consequently, the activity of the TSH-secreting cells of the anterior lobe of the pituitary will have been suppressed via the short, negative, feedback loop (see Figure 48), so these cells no longer react to TRF.

(b) These individuals have very low T_3 and T_4 levels so the secretion of TRF and TSH is not suppressed to the same extent as

in normal or hyperthyroid individuals. (You might wonder why additional TRF was able to stimulate further secretion of TSH in these circumstances. The answer is that the TSH-secreting cells are able to adjust their sensitivity; this point is explained in detail later in Section 8.1.)

(c) The test shows that the anterior pituitary cells respond to TRF; so either (i) the hypothalamic neurosecretory neurons are not secreting TRF, or (ii) the thyroid gland is not responding to TSH.

(d) One could decide where the malfunction was by measuring levels of T_3 and T_4 in the blood *before* and *after* the injection of synthetic TRF, or measuring the level of TRF and comparing it with the level found in normal individuals. High levels of T_3 and T_4 after the injection of TRF, or a low level of TRF, would indicate a fault in the hypothalamus. Unchanged levels of T_3 and T_4 after the injection of TRF, or a normal level of TRF, would indicate that the thyroid gland was not responding to TSH.

SAQ 20 (i) The level of ACTH will be decreased because of negative feedback by the elevated cortisol (Section 7.1). (ii) The ratio of adrenalin to noradrenalin in the blood will change because high levels of corticosteroids stimulate the adrenal medulla to secrete more adrenalin (Section 6.2). (iii) The levels of thyroid-stimulating hormone (TSH) and thyroxin (and triiodothyronine) will increase because cortisol increases the sensitivity of TSH-secreting cells to TRF. As you can see, this single change in the level of one hormone has profound effects on other parts of the endocrine system. Transplant patients must take a whole array of drugs to combat these and other side-effects in addition to the immunosuppressant drugs.

SAQ 21 A *low* K_d value indicates a high affinity, so nuclear receptors have a higher affinity for T_3 than T_4. Because the receptors are more efficient at binding T_3, lower doses of T_3 would be more effective at eliciting a response in the cell.

SAQ 22 Testosterone and progesterone have no effect on the number of TRF receptors. Cortisol *increases* the number of receptors but T_3 and TRF both *reduce* it. Therefore, cortisol *sensitizes* the cells to TRF whereas T_3 and TRF cause desensitization.

SAQ 23 The conclusive test would be to incubate liver cells with each of substances A–E and measure changes in the level of cyclic AMP and protein production in the liver cells. If any one of A–E produces increases in *both*, then it must be an analogue of Z acting on the liver cell receptor. If there is no increase in either the level of cAMP *or* protein production, this suggests the compound does not work in the same way as Z.

SAQ 24 Table 6 shows the results obtained from an experiment such as that described in SAQ 23. The criteria to be met by an *agonist* are that: it increases levels of cAMP and it increases the production of protein. B, C and E all increase the production of the protein (i.e. they are potential agonists), but C produces no change in cAMP. (If the 10 per cent is a meaningful change in protein, C is acting not via cAMP but a different second messenger.) B produces only a 50 per cent rise in protein and cAMP, so although it is an agonist it is a relatively *weak* one compared with E. A produces no changes in either cAMP or protein and D produces a slight rise in cAMP. Neither appear to be agonists but both *could* be antagonists.

The key findings that sort out these five substances are shown in the second half of the Table. Here, all have been added to liver cells in the presence of Z. Remember Z should produce 100 per cent increases in both protein and cAMP. With C and D it does, so *neither* are relevant here. In the presence of B and E, Z stimulates a higher than expected production of protein and level of cAMP. Both B and E therefore add to the Z effect but not in a strictly summative way. The effect of Z in the presence of A is nil. This demonstrates that A is certainly an antagonist, preventing Z from acting by blocking the liver cell receptors.

So the answers are:

A   antagonist

B   'weak' agonist

C   neither agonist nor antagonist

D   neither agonist nor antagonist

E   agonist.

References and further reading

References to the Foundation Course

S101

1 Unit 22, *Physiological Regulation*, Section 3.1.
2 Unit 8, *Energy*, Section 10.
3 Unit 22, Sections 5 and 7.
4 Unit 22, Sections 5 and 7.
5 Unit 22, Section 3.2.

S100

Unit 18, *Cells and Organisms*, Section 18.2.4.
Unit 4, *Forces, Fields and Energy*, Section 4.4.
Unit 18, Section 18.3.3.
Unit 18, Section 18.3.3.
Unit 18, Section 18.4.

Further reading

All of the following are relatively inexpensive, and you should have no difficulty with the level of discussion.

Bently, P. G. (1976) *Comparative Vertebrate Endocrinology*, Cambridge University Press.

Beale, R. and Lagnado, J. R. (1979) 'Macromolecular aspects of communication in the nervous system' in Bull, A. T., Lagnado, J. R., Thomas, J. O. and Tipton, K. F. (eds.) *Companion to Biochemistry*, Vol. 2, Longman.

Bowsher, D. (1979) *Introduction to the Anatomy and Physiology of the Nervous System*, 4th edn., Blackwell.

Cooper, J. R., Bloom, F. E. and Roth, R. H. (1978) *The Biochemical Basis of Neuropharmacology*, Oxford University Press.

Highnam, K. C. and Hill, L. (1977) *The Comparative Endocrinology of the Invertebrates*, 2nd edn., Edward Arnold.

Rose, S. P. R. (1976) *The Conscious Brain*, Pelican.

Acknowledgements

Grateful acknowledgement is made to the following for permission to reproduce material in this Unit:

Figures 7, 8 and 14 modified from Kuffler, J. W. and Nicholls, J. G. (1976) *A Cellular Approach to the Function of the Nervous System*, Sinauer Associates Inc. Copyright © Stephen W. Kuffler and John G. Nicholls; *Figure 34a* modified from Schulster, D. (1976) *Molecular Endocrinology of the Steroid Hormones*, John Wiley. Copyright © John Wiley and Sons Ltd. Reprinted by permission; *Figure 34b* modified from Kreiger, D. T. (1972) Circadian corticosteroid periodicity: critical period for abolition by neonatal injection of corticosteroid, *Science*, **178**. Copyright © 1972 by the American Association for the Advancement of Science; *Figure 38a* modified from Davson, H. and Segal, B. (1975) *Introduction to Physiology*, Academic Press; *Figure 51* from Fletcher, R. F. (1978) *Lecture Notes on Endocrinology*, Blackwell Scientific Publications.

unit 17

Blood Sugar Regulation

Contents

TABLE A Scientific terms and principles used in Unit 17

Assumed knowledge	Introduced in an earlier Unit	Unit	Introduced or developed in this Unit	Page
coma	acetylcholine	16	adrenalin*	16
fat	acetyl-CoA	8	autonomic control of blood glucose level*	16
fat cells	action potential	16	cold stress*	12
liver cells	adrenal gland	16	diabetes mellitus*	20
metabolism	adrenalin	16	enzyme activation and deactivation	5
muscle cells	adrenocorticotropic hormone (ACTH)	16	feedback mechanisms*	4
phenotype		16	glucagon*	12
polymerization	amines	16	glucocorticoids*	16
polypeptide	amino acids	5	gluconeogenesis	6
relative molecular mass	antagonist	16	glucose concentration in the blood*	4
	antibodies	16	glycogenesis	5
	autonomic nervous system (ANS)	16	glycogenolysis	5
	autoradiography	4	glycogen phosphorylase*	5
	β-oxidation of fatty acids	8	glycogen synthase*	5
	calmodulins	16	glycosuria*	20
	Ca^{2+}, cyclic AMP as second messengers	16	homeostasis*	4
	cell junctions	4	hormone	5
	cisternum	4	hyperglycaemia*	4
	cytoplasm	4	hypoglycaemia	4
	endocrine glands	16	insulin*	9
	endoplasmic reticulum (ER)	4	insulin treatment of diabetes*	21
	endorphin	16	interconversion of carbohydrates, fats and proteins*	5
	enkephalin	16	islets of Langerhans*	8
	enzymes	6, 7	islet cells (A, B and D)*	8
	exocytosis	4	juvenile-onset (insulin-dependent) diabetes*	21
	fatty acids	5	keto-acidosis*	13
	feedback mechanisms	7, 16	ketone bodies*	6
	glucocorticoids	16	ketotic coma*	21
	gluconeogenesis	8	maturity-onset (non-insulin-dependent) diabetes*	21
	glucose, glucose phosphates	5, 8	pancreas	8
	glycerol	5	pancreatic peptides	14
	glycogen	5	phosphodiesterase*	5
	glycogen synthase	8	pre-pro-insulin	9
	(glycogen) phosphorylase	8	proglucagon	12
	glycolysis	8	protein kinase*	5
	glycogen mobilization	8	receptor sensitivity to hormones	11
	glycogenesis	8	somatostatin*	14
	Golgi apparatus	4	synergistic effect of hormones on blood glucose*	20
	gonadotropin	16		
	growth hormone	16		
	homeostasis	1		
	hormones	16		
	hypothalamus	16		
	ketone bodies	8		
	lactate	8		
	FSH/LH-RF	16		
	microelectrode	9 & 10		
	mitochondria	4		
	noradrenalin	16		
	osmosis	9 & 10		

* These terms must be thoroughly understood—see Objective 1.

Assumed knowledge	Introduced in an earlier Unit	Unit	Introduced or developed in this Unit	Page
	pancreas	16		
	paracrine cells	16		
	phosphodiesterase	16		
	phosphorylation	8		
	pituitary gland (hypophysis)	16		
	pro-insulin	5		
	protein kinases	8		
	pyruvate	8		
	receptors	16		
	receptor regulation	16		
	second messenger	16		
	signal sequence	5		
	steroid hormones	16		
	sympathetic and parasympathetic nervous systems	16		
	synergism	16		
	substrate	8		
	transcription	16		
	thyroid-stimulating hormone releasing factor (TRF)	16		
	tricarboxylic acid (TCA) cycle	8		

Study guide for Unit 17

Unit 17 is short and represents half a normal week's work. It is concerned with the regulation and control of blood glucose levels, with passing reference to the fate of amino acids and fats in the body. The Unit relies heavily on your knowledge of the metabolism of glucose, amino acids and fatty acids derived from Unit 8, and Section 1 revises the effects of hormones on enzymes. Having studied Unit 16, you will be familiar with many of the hormones and terms that you will come across in this Unit.

After a review of carbohydrate, fat and amino acid metabolism (Section 1), the Unit describes the major hormones produced in the endocrine parts of the pancreas, and considers how their release is controlled by circulating nutrients and hormones in the blood (Section 2). The effects of these hormones on their major target organs is to cause them to store or to provide glucose. Section 3 describes how the nervous system is involved in the regulation of blood sugar via the pancreas, and the many facets of blood sugar regulation are brought together in Section 4. Section 5 looks at one particular aspect of the control of blood glucose: how it goes wrong in diabetes. Sections 2 and 4 are the most important sections in this Unit.

The television programme *Diabetes: restoring the balance* looks at the causes, the treatment and the complications of diabetes, and thus relates strongly to Section 5.

There is reference to *The S202 Picture Book** in Section 2, so have this to hand.

---

* The Open University (1981) S202PB *The S202 Picture Book*, The Open University Press. This contains colour and half-tone illustrations.

1 The uptake and use of glucose

In Section 1.2, we revise concepts introduced in Unit 8, Sections 2.5 and 4.1. Make sure you understand the terms glycogenesis, glycogenolysis and gluconeogenesis and how fats, amino acids and glucose are metabolized.

In Unit 16, we looked at the way that the endocrine and nervous systems function. Working together, these systems regulate and control many physiological processes in the body.

One of the major themes of physiology is *homeostasis*. The cells of an organism live in an environment that is regulated within narrow limits. If the external environment of the organism does not happen to fall within these limits, internal adjustments must be made to the body fluids bathing the cells. A number of factors may shift the composition of the body fluids beyond these limits. Products of cell metabolism will tend to accumulate in the blood, and the amount of oxygen surrounding the cells will vary depending on the activity of the tissue. How oxygen supplies to the tissues are maintained and how metabolites are removed are the subjects of later Units in this Course.

homeostasis*

There are many instances of nerves and hormones playing a crucial part in homeostatic mechanisms, often through complex *feedback mechanisms*. In this Unit, we look in detail at one such mechanism, the regulation of the level of the simple sugar, glucose, in the blood. The level of glucose in the blood is kept within strict limits because it is an essential substrate for cell metabolism. Glucose is added to and removed from the blood very rapidly for a number of reasons, and this calls for an extremely sensitive regulatory mechanism.

feedback mechanisms*

1.1 Glucose levels in the blood

The level of glucose in human blood normally varies between 80 and 110 mg per 100 cm^3. This variation is due to the absorption of food after meals, which raises blood glucose levels, and moderate fasting between meals (particularly overnight), when glucose levels fall. Figure 1 shows such a 24 hour profile of the level of blood glucose. There are two important observations to be made on this Figure.

glucose concentration in the blood*

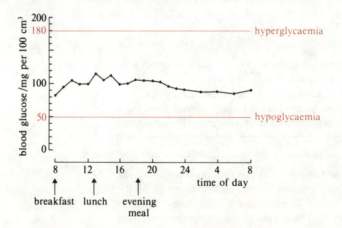

FIGURE 1 Changes in the level of glucose in the blood over a 24 hour period. Note that the levels rarely vary outside 80–110 mg per 100 cm^3.

First, a meal may contain carbohydrate in amounts that would present the blood with about a hundred times the normal level of glucose, and yet the rise in blood glucose is fairly small and transient. Second, an overnight fast of 12–14 hours lowers the blood glucose level, but only by some 20 per cent. This suggests that some mechanism must rapidly clear the glucose absorbed immediately after a meal and that glucose must be made available to the blood from a body 'store' during a fast.

When glucose levels rise or fall beyond these limits, the consequences can be severe. People with blood glucose levels below 50 mg per 100 cm^3, a condition known as *hypoglycaemia*, report feelings of nausea, cold sweats and loss of concentration. If the glucose level is not raised rapidly, coma can result because the nervous system (with no metabolic stores of its own) is uniquely dependent on glucose. If the level of glucose in the blood rises above 180 mg per 100 cm^3, a state of *hyperglycaemia*, the blood becomes very acid, and glucose begins to appear in the urine. Hyperglycaemia can also result in coma, which, if unchecked,

hypoglycaemia

hyperglycaemia*

4

is fatal. These hypoglycaemic and hyperglycaemic states are typical of people who have problems regulating their blood glucose level; their condition, known as diabetes mellitus, is considered in Section 5.

1.2   The interconversion of carbohydrates, fats and proteins

To understand the events during eating and fasting, it is important to remember that cells, particularly liver cells, have a remarkable ability to convert one type of molecule into another. The *interconversion of carbohydrates, fats and proteins* depends on a complex series of events catalysed by enzymes in liver cells, first described in Unit 8, Section 4. Before we look at the role of the endocrine and nervous systems in regulating the levels of blood glucose, we need to revise our knowledge of these interconversion processes and see how the body stores glucose.

interconversion of carbohydrates, fats and proteins*

Glucose is converted to glycogen in liver and muscle cells. Glycogen can be conveniently stored by cells and easily broken down again to glucose.

□ What are the main steps in the conversion of glucose to glycogen?

■ Glucose is first phosphorylated, then activated and then polymerized to form glycogen (Unit 8, Section 4.1.5).

The polymerization step is catalysed by the enzyme *glycogen synthase*.

glycogen synthase*

□ What are the main steps in *glycogenolysis* (the breakdown of glycogen to glucose)?

glycogenolysis

■ Glycogen is phosphorylated to glucose 1-phosphate, which is isomerized to glucose 6-phosphate, and converted to glucose by the enzyme glucose 6-phosphatase (Unit 8, Sections 2.1 and 4.1.3).

The phosphorylation reaction is catalysed by the enzyme *glycogen phosphorylase*.

glycogen phosphorylase*

The stimulation or inhibition of these two enzymes, glycogen synthase and glycogen phosphorylase, is by the interconversion of their two forms (*a* and *b*), one of which is active (*a*) and the other inactive (*b*). The conversion of the *b* form (inactive) to the *a* form (active) of glycogen phosphorylase stimulates the breakdown of glycogen. Similarly, conversion of the *b* form (inactive) of the glycogen synthase to the *a* form (active) promotes the synthesis of glycogen (*glycogenesis*). The key to the interconversion of these *a* and *b* forms of the two enzymes is protein phosphorylation dependent on cyclic AMP (Unit 8, Section 3.3).

enzyme activation and deactivation

glycogenesis

□ How might cyclic AMP be involved?

■ From Unit 16, Section 8, you will remember that cyclic AMP is involved in the activation of a number of *protein kinases* in cells.

protein kinase*

The inactivation (*a* → *b*) of glycogen synthase in liver is catalysed by a synthase kinase dependent on cyclic AMP and the very same kinase activates (*b* → *a*) glycogen phosphorylase. So the production of one kinase *switches off* glycogen production and *switches on* glycogen breakdown simultaneously. Because cyclic AMP production plays a part, it will come as no surprise that *hormones* are involved at this stage. One of the main hormones in question is glucagon, and we return to it in Section 2.2.

Conversely, if cyclic AMP production is blocked or its breakdown is stimulated (by *phosphodiesterase*), or if the activation of the protein kinase by cyclic AMP is inhibited, then any one or all of these events would tend to activate the synthesis of glycogen and to prevent its breakdown. It is at this point that another important hormone, insulin, is involved (see Section 2.1).

Glucose itself has an effect on this process of glucose–glycogen conversion. Glucose inhibits the *a* form of glycogen phosphorylase. In Figure 2, the levels of phosphorylase *a* and synthase *a* measured in rats after a glucose meal are shown. The level of phosphorylase *a* is seen to drop rapidly and then synthase *a* rises.

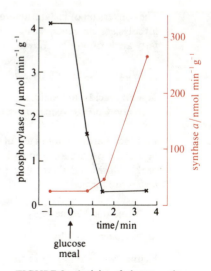

FIGURE 2   Activity of glycogen phosphorylase *a* and glycogen synthase *a* in the liver after giving a glucose meal to a rat.

□ What will the consequence of these changes be in terms of glycogen?

■ Glycogenolysis will stop and glycogen synthesis will be promoted, leading to glycogen build-up.

5

It is thought that, above a certain level, phosphorylase *a* inhibits synthase *a* formation. The presence of high levels of glucose in liver causes the conversion of phosphorylase *a* to phosphorylase *b* (by blocking the reaction *b* → *a*), and so synthase *a* is no longer inhibited. The effects produced by cyclic AMP and glucose on liver enzymes are summarized in Figure 3.

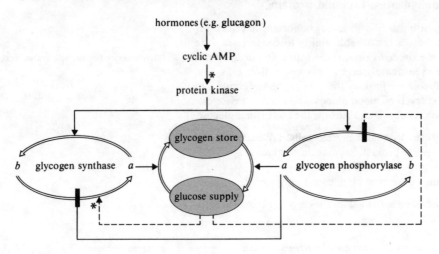

hormones (e.g. glucagon)

cyclic AMP

protein kinase

b   glycogen synthase   *a*

glycogen store

glucose supply

a   glycogen phosphorylase   *b*

FIGURE 3   The control exercised by cyclic AMP and glucose over the inter-conversion of the two forms of glycogen synthase and of phosphorylase in the liver. In this and other figures in this Unit, a line ending in a thick bar indicates inhibition, and a line ending with an arrow and a star indicates stimulation. Feedback loops are dashed.

A similar relation between production and breakdown has been found in the other major glucose store, muscle glycogen.

Muscle cells, though, lack the enzyme glucose 6-phosphatase.

☐ What is the significance of this enzyme deficiency?

■ The breakdown of glycogen in muscle does not produce glucose, but glucose 6-phosphate, which is directed down the glycolytic pathway and then metabolized via the TCA cycle.

Glucose is made available from liver glycogen stores for distribution to other body tissues when the levels of blood glucose fall. But liver glycogen is not the only source of glucose.

☐ What are the main routes of *gluconeogenesis*?

                    gluconeogenesis

■ 1   Lactate (from anaerobic respiration) through pyruvate (Unit 8, Section 4.1.4).
2   Alanine (from protein breakdown) through transamination to pyruvate (Unit 8, Section 4.1.4).
3   Glycerol (from the breakdown of fats) through pyruvate (Unit 8, Sections 2.5.1, 4.1.4).

The conversion of lactate, alanine and glycerol to glucose is achieved by reversed glycolysis. Remember, though, from Unit 8, Section 4.1.4, that in most mammals fatty acids cannot be converted to glucose.

☐ What happens to fatty acids?

■ Fatty acids are oxidized in mitochondria (by β-oxidation) and provide acetyl-CoA for oxidation via the TCA cycle (Unit 8, Section 2.5.2).

Fatty acids are therefore a useful substrate for cellular respiration, but if high levels of acetyl-CoA are allowed to build up (e.g. during starvation when fat is being broken down and there is no production of oxaloacetate from glucose), *ketone bodies* are formed.

                    ketone bodies*

☐ Name one ketone body.

■ Acetoacetate, β-hydroxybutyrate or acetone (Unit 8, Section 2.6).

Ketone bodies build up in the blood in severe diabetes and can cause coma (as we see in Section 5).

The glucose circulating in the blood can thus be derived from a number of sources:

1 from the intestine

2 from the breakdown of liver glycogen

3 from the metabolism of amino acids, lactate and glycerol in the liver (gluconeogenesis).

The interconversion of carbohydrates, fats and proteins is summarized in Figure 4.

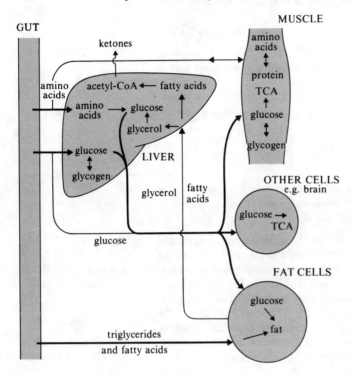

FIGURE 4 The relationship between carbohydrates, fats and protein, and their movement between gut, liver, muscle, fat cells and other cells. The quantitatively more important routes are shown as bold arrows.

Summary of Section 1

1 Glucose levels in the blood are maintained within strict limits, rising after meals and falling overnight or after a period of fasting only slightly.

2 When glucose levels rise or fall beyond these limits, hyperglycaemia or hypoglycaemia results, and there is a danger of coma.

3 Glucose is stored as glycogen in liver and muscle cells. When the levels of blood glucose fall, liver glycogen is broken down to glucose.

4 Fasting depletes glycogen stores and stimulates gluconeogenesis from glycerol and amino acids.

5 The enzymes responsible for the synthesis and breakdown of glycogen (glycogen synthase and glycogen phosphorylase) exist in an active (a) and an inactive (b) form. The conversion of these forms is sensitive to cyclic AMP and glucose. Hormones are involved in the interconversion of glucose and glycogen via cyclic AMP.

Objectives and SAQ for Section 1

Now that you have completed this Section, you should be able to:

★ describe, with the aid of a diagram, how blood glucose levels change over a 24 hour period.

★ list the enzymic steps in the interconversion of glucose and glycogen.

★ state three routes of gluconeogenesis.

To test your understanding of this Section, try the following SAQ.

SAQ 1 (*Objective 2*)   Select two statements from (i)–(v) that you consider to be accurate.

(i)   Glycogen phosphorylase and glycogen synthase exist in two forms (*a* and *b*). In both cases, conversion from *b* to *a* is dependent on the activation of a protein kinase dependent on cyclic AMP.

(ii)   Fatty acids, glycerol and amino acids are all substrates for gluconeogenesis.

(iii)   In the absence of oxaloacetate, the oxidation of fatty acids will result in the production of ketone bodies.

(iv)   Stimulation of a phosphodiesterase that breaks down cyclic AMP inhibits the activation of phosphorylase.

(v)   As the level of glucose is increased, liver cells show a progressive loss of glycogen synthase *a*.

2   Pancreatic hormones

In this very important Section, we describe the release and mode of action of the main glucose regulating hormones, insulin and glucagon. Figures 10 and 11 summarize the major effects of these hormones on target tissues.

Following digestion, the products are absorbed by the small intestine and carried directly to the liver, before being carried to the heart and round the rest of the circulation. Situated just below the liver and under the stomach is the *pancreas* (Unit 16, Section 6). The pancreas plays an important role in the digestion and assimilation of food. 98 per cent of the pancreatic cells produce digestive enzymes. (You will learn more about these in Units 24 and 25). The other 2 per cent of the cells are arranged in discrete 'islands' among the mass of the pancreas, called the *islets of Langerhans*. These islets are composed of a number of cell types, and each produces and secretes a particular hormone.

pancreas

islets of Langerhans*
See *The S202 Picture Book* for pancreas.

If a glucose meal is given to an animal from which the pancreas has been removed, there is a dramatic increase in the level of glucose in the blood. When the glucose level rises above the threshold for the kidney, glucose appears in the urine: glucose homeostasis has been lost! This suggests that the pancreas secretes a hormone that normally *lowers* the blood glucose level. But that is not the end of the story. The islets of Langerhans produce a number of substances of which three—insulin, glucagon and somatostatin—are known to be important in glucose regulation.

By using specific histological staining techniques, it is possible to 'map' the cells within the islets. A number of cell types can be identified. The three main cell types are called A cells, which contain a hormone called *glucagon*, B cells, which contain the hormone *insulin* and D cells, which contain *somatostatin* (growth hormone release inhibiting factor). (In some texts and in the TV programme on diabetes, you may find these designated as α, β and D cells.) This is illustrated in Figure 5. Note that cells are arranged in a distinct pattern in the rat, with

islet cells (A, B and D)*

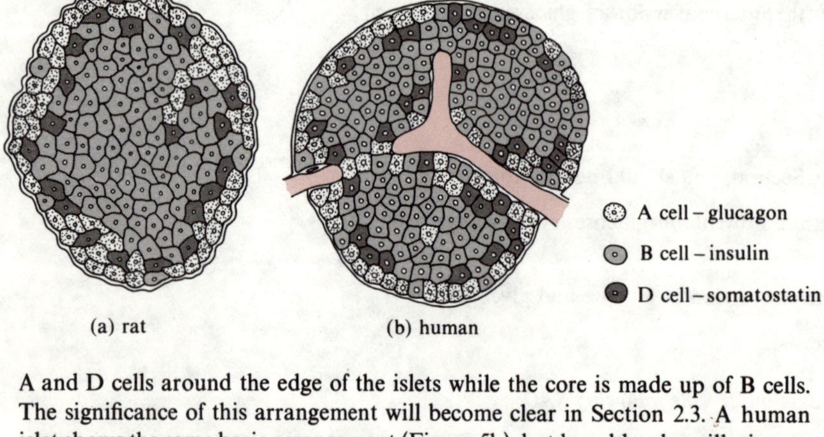

⊙ A cell – glucagon
◉ B cell – insulin
● D cell – somatostatin

(a) rat   (b) human

FIGURE 5   Arrangement of cells in pancreatic islets from (a) rat and (b) human.

A and D cells around the edge of the islets while the core is made up of B cells. The significance of this arrangement will become clear in Section 2.3. A human islet shows the same basic arrangement (Figure 5b), but here blood capillaries are seen to run through the tissue.

□ What difference is there in the arrangement of A, B and D cells in the human islet compared with rats?

■ Because the A and D cells lie along the capillaries, they are found deeper in the islet than in the rat; they appear scattered throughout the islet.

To understand the role of these pancreatic hormones in blood glucose regulation, we need to look at the *synthesis*, *release* and *actions* of each one in turn. In Section 2.3, we return to the morphology of the islet to see how A, B and D cells interact with one another.

2.1   Insulin

B cells make up about 60 per cent of the islet mass. They synthesize and release the hormone *insulin*. Insulin is a peptide of 51 amino acids, but is derived from a larger parent molecule called *pre-pro-insulin* (some 110 amino acids long).

insulin*
pre-pro-insulin

□ On the route to secretion from the cell, pre-pro-insulin loses a hydrophobic, 23 amino acid, peptide and becomes pro-insulin inside the endoplasmic reticulum. What might the function of this loss be? (Think back to protein insertion, Unit 5, Section 8.)

■ The hydrophobic part of pre-pro-insulin is probably the signal sequence of amino acids that directs pro-insulin through the endoplasmic reticulum (ER) membrane into the ER cisternum. Once inside the ER, this signal sequence is 'clipped' off.

Figure 6 shows the fate of pro-insulin. It is packaged in the Golgi apparatus, and then post-synthetic processing occurs (see Unit 16, Section 6) by enzymic action, and it is clipped into two peptides, one peptide being insulin and the residue being known as the connecting peptide (C-peptide).

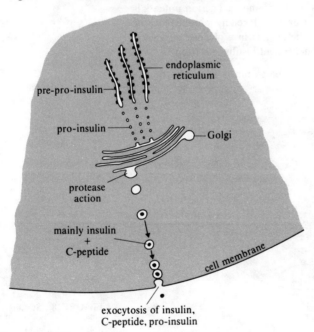

FIGURE 6   Steps in the synthesis of insulin in the B cell.

□ Can you recall any other chemical messenger molecules produced in this way?

■ A number of peptides are produced from parent molecules, for example ACTH, the endorphins and enkephalins (see Unit 16, Sections 6 and 7).

Insulin is stored in granules bound to zinc and released by exocytosis (Unit 4, Section 4.5).

One important stimulus for insulin secretion is an increase in the level of glucose in the blood flowing through the pancreatic islets, for example after a meal. Figure 7 shows the dramatic effect of raising the external glucose concentration on insulin release from islets from an isolated pancreas.

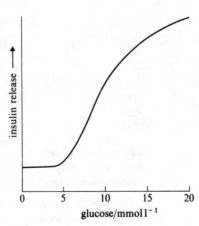

FIGURE 7   The effect of raising the glucose concentration on the amount of insulin released from the islets of an isolated pancreas. ($5 \, \text{mmol} \, l^{-1}$ is equivalent to $90 \, \text{mg}$ per $100 \, \text{cm}^3$ of blood.)

Gut hormones, released into the blood when food is present in the stomach, will also stimulate insulin release *before* the blood glucose level is raised. The key questions are: How does the B cell 'monitor' blood glucose levels? What then stimulates insulin release? If islets are isolated from a pancreas and maintained under Ca^{2+}-free conditions and then exposed to glucose, insulin secretion is reduced, dramatically. If a microelectrode is inserted into a B cell that is then exposed to glucose, an action potential can be recorded. When Ca^{2+} is removed and the experiment repeated, no action potential is recorded. Let us consider three more observations of the response of B cells to glucose. First, they show a rise in cyclic AMP levels; second, they accumulate Ca^{2+} when exposed to glucose; and third, B cells contain calmodulins (Ca^{2+}-binding proteins, see Unit 16, Section 8).

At first, this evidence might appear easy enough to incorporate into a mechanism by which glucose might stimulate insulin release.

☐ From your knowledge of Unit 16 (Section 8), suggest a *three* step sequence between the arrival of glucose at the B cell and the activation of insulin secretion.

■ 1 Activation of Ca^{2+} channels and influx of Ca^{2+} into the cell.

2 Ca^{2+} binds to calmodulin.

3 Calmodulin activates a receptor protein involved in insulin secretion.

This looks simple enough, but one piece of our evidence suggests that glucose *also* stimulates cyclic AMP production in B cells! It now seems that the activation of calmodulin by glucose (via Ca^{2+}) has two effects on B cells—*directly* on insulin secretion and *indirectly* on insulin secretion through cyclic AMP. A similar complementary involvement of Ca^{2+} and cyclic AMP is also seen in the adrenal medullary cell (Unit 16, Section 8).

Exactly what the mechanism is that triggers the Ca^{2+}-induced action potentials remains elusive. Evidence suggests that it is not simply a B cell membrane receptor for glucose itself, but the trigger is probably the change in a glucose *metabolite* inside the B cell and this initiates the Ca^{2+}-dependent insulin secretion.

Figure 8 summarizes these effects of glucose on the secretion of insulin by the B cell.

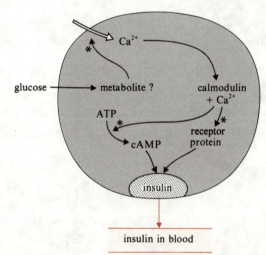

FIGURE 8   Stimulation of insulin release from the B cell by glucose. Note that both Ca^{2+} and cyclic AMP are involved.

Glucose is not the only stimulus for insulin secretion. Other sugars, amino acids, ketones and a range of hormones (including those from the gut) initiate insulin secretion from the pancreas.

Insulin circulates in the blood throughout the body, and receptors for insulin have been found on three main target organs: liver cells, fat cells and muscle tissue. The main effect of insulin is to lower the level of glucose in the blood by stimulating glycogenesis (in the liver and in muscle) and by inhibiting gluconeogenesis in the liver. Insulin has a number of other general effects: it promotes the uptake of amino acids into cells and prevents the breakdown of fat and protein in tissues such as fat cells and muscle.

Earlier, in Section 1.2, we saw how glucose in the liver is converted to glycogen, which can then be broken down again to glucose when required.

☐ What are the two key enzymes?

■ Glycogen synthase (glucose → glycogen) and glycogen phosphorylase (glycogen → glucose).

Generally, the actions of insulin are best thought of in terms of *promoting the build-up* and *preventing the breakdown* of fat, protein and glycogen stores. This is achieved by quite a complex series of events, which we will now try to unravel.

First, we consider glycogen storage, because this has been most studied. In the absence of insulin, we know that glucose on its own will activate glycogen *synthase* (see Section 1.2), and in the absence of glucose, insulin alone will activate glycogen synthase. In *combination*, insulin and glucose produce a synthase activation response *greater* than the sum of each separately—a synergistic effect. This suggests that *two* separate mechanisms are at work—insulin promotes glucose uptake by tissues, and this alone will promote synthase activity and the production of glycogen; the other effect of insulin appears to be a more direct one on the synthase enzyme itself.

You will recall from Section 1.2 that glycogen synthase exists in two forms, an active (*a*) form and an inactive (*b*) form. The *a* → *b* conversion is dependent on a protein kinase reaction induced by cyclic AMP. Insulin seems to *prevent* this reaction and so maintains synthase in the *a* or active form (see Figure 9). The result is that glycogen is deposited in the liver. Insulin could achieve this in two ways: it could stimulate the breakdown of cyclic AMP (through the enzyme phosphodiesterase—see Unit 16, Section 8), or it could stimulate the production of a protein kinase inhibitor. At present (1980), evidence for *both* mechanisms exists! Insulin will *inhibit the activation of cyclic AMP* itself (produced by other hormones) in liver and fat cells. Equally, insulin appears to *stimulate the production of a protein kinase inhibitor* and so block the *a* → *b* conversion of glycogen synthase by cyclic AMP.

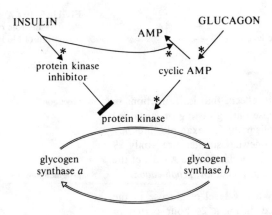

FIGURE 9 The effects on the inter-conversion of glycogen synthase *a* and *b* forms by insulin and gluconeogenic hormones (e.g. glucagon).

The key to understanding the cellular actions of insulin lies in discovering exactly what second messenger the insulin receptor 'uses' to achieve its effects. So far, this link has remained elusive.

Before we leave insulin, there is one important observation to be made about its receptor. In conditions such as obesity, the level of insulin and the level of glucose in the blood are *both* very high. This apparent contradiction is due to the relative *insensitivity* of tissues to insulin. Insulin receptors are present, but are very few in number, and it seems that the constant presence of the hormone actually reduces the number of its receptors. Reduced '*sensitivity*' *of receptors to hormones* has been described for a number of hormones (see Unit 16, Section 8). In all other ways, the hormone receptor event is normal; it is simply that there are fewer such events. When obese animals are fasted to reduce the obesity and therefore the levels of insulin in the blood, the number of insulin receptors increases again. We return to the concept of receptor regulation in Section 5 when describing the problems of glucose regulation in some diabetics. To summarize this Section on insulin, look at Figure 10 (overleaf), and concentrate particularly on the actions of insulin.

receptor sensitivity to hormones

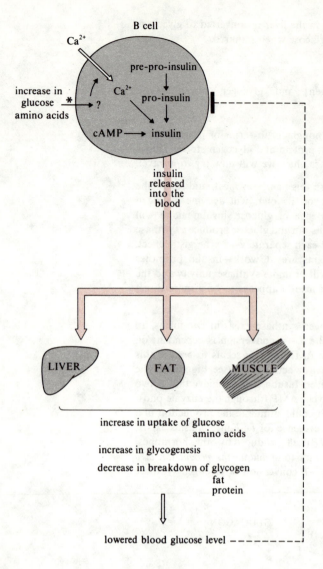

increase in uptake of glucose
amino acids

increase in glycogenesis

decrease in breakdown of glycogen
fat
protein

lowered blood glucose level – – – – – – – –

FIGURE 10 A simplified scheme showing the role of insulin in the regulation of blood glucose.

2.2 Glucagon

Glucagon is a hormone *antagonistic* to insulin in its effects, that is, it functions in ways that lead to increased levels of blood glucose during fasting or sustained exercise when glucose is in demand for metabolism in, for example, muscles. Like insulin, glucagon is a polypeptide but of somewhat smaller size—only 29 amino acids (relative molecular mass 3 500). It is synthesized in the A cells of the pancreas (which form 25% of the islets cells) from a precursor, *proglucagon*.

glucagon*

proglucagon

The stimulus for glucagon release is not simply a low level of glucose in the blood. If circulating blood glucagon is measured during a 24 hour period in healthy subjects (eating mixed meals), there is little change in its level. Stimulation of a marked glucagon release from the pancreas requires a definite 'stress'—for example, brief starvation or an absolute lack of insulin.

The islets of Langerhans are arranged (Figure 5) such that there is considerable communication between the A, B and D cells (by way of cell junctions and extracellular spaces). Insulin release inhibits the release of glucagon, and in the absence of insulin, glucagon is released. This fine feedback control of each other's secretion ensures that glycogen is stored in times of excess blood glucose. In times of glucose demand by tissues (e.g. in exercise, or during increased metabolism in response to a *cold stress*), the circulating insulin level is low and release of glucagon is stimulated to 'pull' glucose out of store and into the blood. Blood glucose is regulated therefore by two hormones—insulin and glucagon.

cold stress*

The major site of glucagon action is on liver cells, where it promotes the breakdown of glycogen by activating the enzyme glycogen phosphorylase (Section 1.2) and also blocks the production of glycogen by deactivating the enzyme glycogen synthase. Unlike insulin, glucagon appears to act solely via cyclic AMP. Cyclic AMP activates protein kinase, which in turn promotes glycogen phosphorylase activity $(b \rightarrow a)$ and 'turns off' glycogen synthase $(a \rightarrow b)$. The result is that glycogen stores are broken down.

In the absence of insulin, fat and protein will break down, liberating fatty acids, glycerol and amino acids. Glucagon stimulates the uptake of amino acids into the liver, where they are converted to glucose (Section 1.2). Fatty acids and glycerol are metabolized by the liver to provide acetyl-CoA and pyruvate, respectively (Section 1.2). Glucagon also promotes gluconeogenesis through direct effects on liver mitochondrial enzymes.

Glucagon, like insulin, appears to affect the density of its receptors on liver cells. In prolonged fasting conditions when glucagon levels are high, there is a desensitization of liver cells to glucagon. The physiological significance of this receptor regulation may be that it is an 'escape response' that prevents *continuous* breakdown of fat and structural protein.

☐ What would the consequence be of continual fat and protein breakdown?

■ Fat breakdown liberates fatty acids, which are converted to acetyl-CoA, and if in excess, to ketone bodies (i.e. *keto-acidosis* could result). Prolonged protein breakdown could result in severe wastage of the muscles.

keto-acidosis*

Figure 11 summarizes the main effects of glucagon. The major stimulus for glucagon release from the pancreas is the *lack* of insulin. Other stimuli (such as stress, starvation and exercise) promote glucagon release from the pancreas through other pathways, as we see in the next Section.

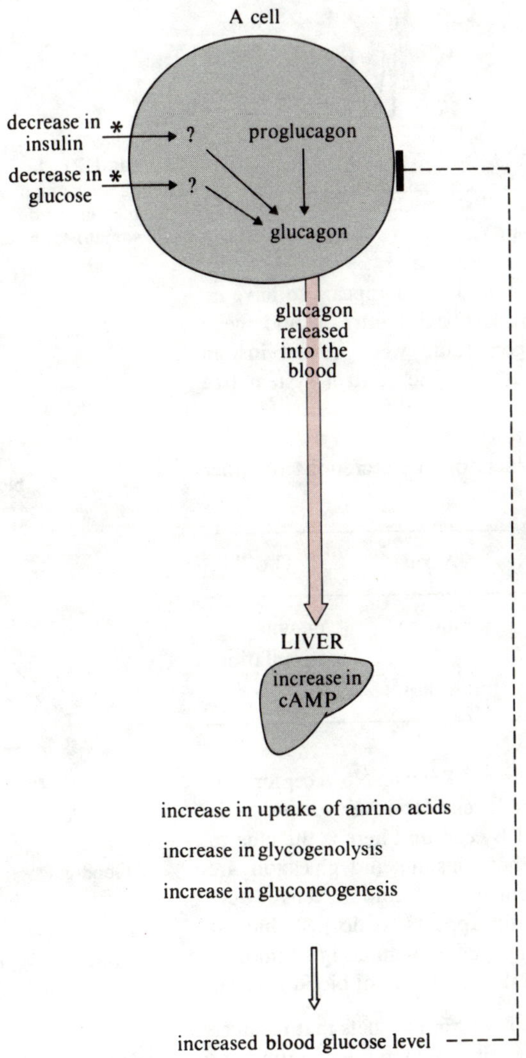

FIGURE 11   A simplified scheme showing the role of glucagon in the regulation of blood glucose.

2.3 Somatostatin and other pancreatic peptides

Insulin and glucagon acting in concert maintain a fairly constant blood glucose level throughout the day (see Figure 1), and if we were to measure the ratio of these two hormones in the blood at any one moment, it would correlate well with the level of blood glucose.

In Section 2, we described the presence of a third cell type, which makes up 10 per cent of the islets of Langerhans—the D cell. Description of pancreatic endocrine function before 1975 put a large question mark against the D cell. Following the isolation of somatostatin from the hypothalamus, and therefore the possibility of raising antibodies to it for the purpose of identification (Unit 16, Section 6), it soon became clear that D cells secrete *somatostatin*.

□ What is the function of somatostatin released from the hypothalamus?

■ Classically, it is thought of as an inhibitor of growth hormone (GH) release from the pituitary (somatostatin is the growth hormone release-inhibiting factor).

So what could its function in the pancreas be? Somatostatin appears to inhibit, locally, the release of *both* insulin and glucagon, and the anatomical position of these D cells (Figure 5) puts them in a perfect position to achieve this effect. In other words, somatostatin has a paracrine function (Unit 16, Section 7). Glucagon release from the A cell stimulates insulin release through activating cyclic AMP in B cells and also promotes the secretion of somatostatin from the D cells (see Figure 12). Here then is a very tight, negative feedback loop between glucagon

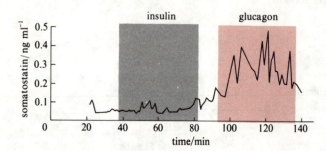

FIGURE 12 The effect of giving the pancreas high doses of first insulin and then glucagon on the amount of somatostatin released.

and somatostatin, and between glucagon and insulin. Insulin appears to have no effect on somatostatin (Figure 12). The negative feedback control would mean that somatostatin release mediated by glucagon could well prevent insulin secretion mediated by glucagon. The result is a very fine control system (see Table 1).

TABLE 1 Actions of islet cell secretory products on the secretions of other islet cells

Cells	Secretory product	B cell	A cell	D cell
B	insulin	—	inhibition	no effect
A	glucagon	stimulation	—	stimulation
D	somatostatin	inhibition	inhibition	—

Exactly how somatostatin achieves these effects is not known. No receptors for somatostatin have been found on pancreatic A and B cells. However, somatostatin does appear to affect Ca^{2+} movements in A and B cells, and here is the clue to how it might operate. Because the release of both insulin and glucagon are dependent on Ca^{2+}, any block of Ca^{2+} movement in A and B cells would inhibit insulin and glucagon release. Somatostatin appears to do just that—it blocks the Ca^{2+}-stimulated secretion of the main glucose-regulating hormones. Figure 13 summarizes the role of somatostatin in the regulation of blood glucose.

In addition to A, B and D cells, there are other cells within the islets that produce peptides. A number of such peptides have been identified by the use of antibodies (Unit 16, Section 6) and they are a very mixed collection indeed! Pancreatic islets, in addition to producing insulin, glucagon and somatostatin, produce thyroid stimulating hormone releasing factor (TRF), luteinizing hormone (or gonado-

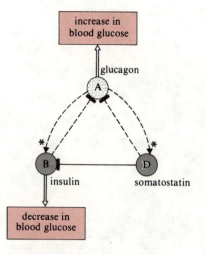

FIGURE 13 A summary of the interactions between insulin, glucagon and somatostatin in the pancreatic islets.

14

tropin) releasing factor, cholecystokinin, and a small peptide designated 'pancreatic polypeptide'. Whether these peptides act locally within the islets on A, B, and D cells, or interact with the rest of the pancreas that produces digestive enzymes, or are released into the blood stream and have more general actions on other tissues remains to be seen.

Summary of Section 2

1   The pancreas contains discrete groups of endocrine cells called the islets of Langerhans. A number of cell types can be identified in the islet, namely the A, B and D cells, which produce and secrete the hormones glucagon, insulin and somatostatin, respectively.

2   The release of insulin is promoted by raised blood glucose levels, which stimulates Ca^{2+} movement in B cells. Ca^{2+} activates a calmodulin-dependent receptor protein, and this is responsible for insulin secretion and also for a rise in cyclic AMP that amplifies the secretion of insulin. Other sugars, together with amino acids and a number of chemical messengers, all stimulate insulin secretion.

3   Insulin promotes glucose and amino acid uptake into cells, prevents the de-activation of glycogen synthase by cyclic AMP, and so promotes glycogen storage. Receptors for insulin have been found on many cell types, and these become desensitized in the presence of a persistently high level of insulin.

4   Glucagon release is promoted by the absence of insulin, by fasting and by prolonged exercise. Glucagon raises blood glucose levels by activating glycogen phosphorylase and deactivating glycogen synthase; it also stimulates gluconeogenesis by increasing amino acid transport into cells and through effects on mitochondrial enzymes. Glucagon receptors also become less sensitive if glucagon levels are high.

5   Somatostatin inhibits glucagon and insulin release from the A and B cells. Somatostatin release is stimulated by glucagon. Somatostatin regulates blood glucose levels by local effects (a paracrine function) on the hormones released from adjacent islet cells.

6   A number of other peptides are also found in pancreatic islets, but their function in the pancreas is unknown.

Objectives and SAQs for Section 2

Now that you have completed this Section, you should be able to:

★ contrast the effects of glucagon and insulin on liver cell glycogen.

★ describe, with a flow diagram, how insulin, glucagon and somatostatin affect each other's secretion.

★ predict what would happen to the blood glucose level if the ability of the pancreas to secrete or respond to glucagon and insulin was diminished.

★ describe the effects of glucagon and insulin on protein and fat stores.

To test your understanding of this Section, try the following SAQs.

SAQ 2 (*Objectives 1 and 3*)   What would the effect of the following be on pancreatic insulin secretion in the presence of glucose?
(a)   Inhibition of somatostatin release from the pancreas.
(b)   Completely removing Ca^{2+} from the pancreas.
(c)   Inhibiting phosphodiesterase activity in B cells.
(d)   The presence of food in the stomach and intestine.
(e)   Blocking glucose transport across B cell membranes.

SAQ 3 (*Objectives 3 and 4*)   Select two conditions from (i)–(iv) under which fat and protein stores would be depleted.
(i)   Lack of B cells in the pancreas.
(ii)   Persistently high levels of blood glucagon.
(iii)   A meal rich in carbohydrate.
(iv)   Blocking somatostatin release.

3 Other control systems in glucose regulation

The nervous system and hormones other than insulin and glucagon have effects on the levels of glucose in the blood. Their mode of action is either through the control of hormone release from the pancreatic islet cells or through direct effects on tissue glycogen stores, or both.

3.1 The autonomic control of blood glucose

So far, we have seen how blood levels of glucose are maintained through the concerted actions of the liver, intestine and hormones from the pancreas. Because the brain requires a continuous supply of glucose for continued function (some 6 g per hour in humans), it would be somewhat surprising not to identify a glucose control mechanism mediated by the nervous system. Similarly, exercise, cold stress and fasting present the body with a 'challenge' to the level of blood glucose.

In the islets of Langerhans can be seen numerous nerve endings from the autonomic nervous system (ANS) that lie close to the A, B and D cells.

☐ What chemical messenger would you expect to find in these nerve endings, remembering that it is ANS innervation?

■ Amines—for example, acetylcholine (from the parasympathetic system) and noradrenalin (from the sympathetic system).

A transplanted pancreas, or a pancreas deprived of nervous input, will still maintain blood glucose levels. These observations have, until recently, dampened any interest in the possible significance of nervous control mechanisms. Both acetylcholine and noradrenalin have been shown to alter the secretion of insulin and glucagon. Noradrenalin inhibits insulin release and stimulates glucagon release. Conversely, acetylcholine increases both insulin and glucagon release, possibly by inhibition of somatostatin release from the D cell.

What physiological significance could these observations have? First, they might represent another level of control over blood glucose, directed by the hypothalamus. There are insulin receptors in the hypothalamus, and also glucose receptors, so that direct regulation could be exerted via a feedback mechanism from the hypothalamus through the parasympathetic and sympathetic nerve fibres to the pancreas. The hypothalamus is a key area involved with temperature regulation and feeding, and thus could command changes in glucose use (in response to cold) or in glucose storage (in response to fasting) by immediate regulation of insulin and glucagon release from the pancreas. The observation that anticipation of a meal and hunger will both elicit insulin secretion probably has a basis in nerve signals from the hypothalamus to the pancreas via the parasympathetic system. If these nerves are cut, the rise in insulin level stimulated by the anticipation of food is not observed.

autonomic control of blood glucose level*

Stress states are also associated with activation of the autonomic nervous system, for example exercise, trauma, surgery, burns, pain, anxiety and infection. All of these conditions can lead to a lowering of the insulin : glucagon ratio, which then results in the mobilization of the body's glucose and fat stores. In such ways, the nervous system can regulate the insulin and glucagon response, overriding control at lower levels, and so promote the metabolism of glucose and the input of glucose to the blood to serve the particular requirements of the nervous system.

3.2 Adrenalin and glucocorticoids

In a less direct way, any discharge of the sympathetic nervous system will also increase the release of adrenalin from the adrenal glands (Unit 16, Section 6). Increased circulating levels of *adrenalin*, like noradrenalin, will inhibit insulin release from the pancreas and promote the release of glucagon. Increased secretion of adrenocorticotropic hormone (ACTH) from the anterior pituitary is also associated with stress via the release of corticotropin releasing factor (CRF) from the hypothalamus, and the resulting rise in *glucocorticoid* secretion from the adrenal cortex affects the pancreas, leading to increased secretion of glucagon (see Figure 14).

adrenalin*

glucocorticoids*

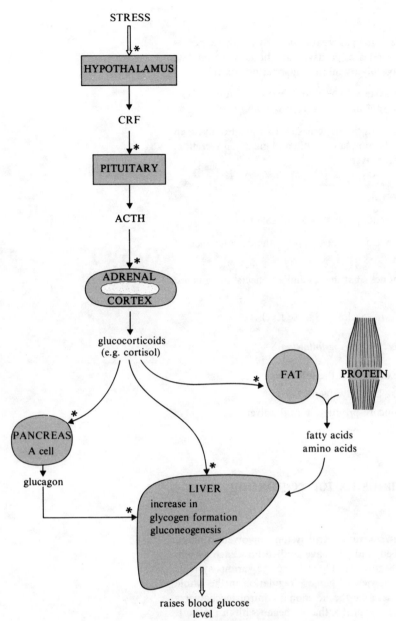

FIGURE 14   The sequence of events whereby stress (through the actions of gluco-corticoids) can mobilize glucose for metabolism at the expense of fat and protein.

Adrenalin and glucocorticoids have additional effects on the liver cells. Adrenalin rapidly activates glycogen phosphorylase ($b \rightarrow a$) and promotes glycogen break-down. In human liver cells, adrenalin works in the same way as glucagon, which, you will remember, also activates glycogen phosphorylase via cyclic AMP. In rats, however, adrenalin appears to work through Ca^{2+} as the second messenger. Adrenalin also mobilizes glycogen stores in muscle, and again like glucagon, adrenalin promotes gluconeogenesis in the liver through activation of mito-chondrial enzymes (Section 2.2).

Glucocorticoids raise blood glucose, fatty acid and amino acid levels. In liver cells, glucocorticoids deactivate glycogen phosphorylase ($a \rightarrow b$) and so preserve glycogen stores. In other tissues, glucocorticoids promote protein breakdown, which releases amino acids. Amino acids stimulate insulin release from the pancreas, which, in turn, maintains glycogen stores. The combined effects of glucocorticoids is to promote glycogen deposition in the liver (see Figure 14).

☐ What is the amplification mechanism induced in the tissues by glucocorticoids?

■ Remember from Unit 16, Section 8, that steroids such as glucocorticoids react with receptors in the cell cytoplasm, which then initiate transcription in the nucleus. The result is the synthesis of specific proteins in the cytoplasm. Neither cyclic AMP nor Ca^{2+} are involved as second messengers.

Summary of Section 3

1 Autonomic nerve fibres terminate in the pancreatic islets. Sympathetic nerves inhibit insulin and stimulate glucagon release; parasympathetic nerves stimulate both glucagon and insulin release possibly by inhibiting somatostatin release.

2 Through the autonomic nervous system (ANS), anticipation of food, stress, or exercise either promotes glucose storage or increases the availability of glucose.

3 Adrenalin inhibits insulin release from the pancreas and stimulates glycogen breakdown in the liver. Glucocorticoids stimulate insulin and glucagon secretion, and promote glycogen formation in the liver.

Objectives and SAQ for Section 3

Now that you have completed this Section, you should be able to:

★ give examples of hormones that act antagonistically in the maintenance of glucose homeostasis;

★ give examples of different hormones that have similar effects on glucose homeostasis.

★ describe how stress or trauma can raise blood glucose levels.

To test your understanding of this Section, try the following SAQ.

SAQ 4 (*Objectives 2 and 4*)   Give examples of the following:
(a) two hormones that directly raise liver glycogen levels;
(b) two hormones that stimulate glucose production in the liver;
(c) two hormones that promote protein and fat metabolism.

4 The integration of mechanisms for regulating blood glucose

In the first three Sections of this Unit, several control systems involved in blood glucose regulation have been described, and you have studied how each system works separately. In this important Section, we try to bring the various systems together and so produce an integrated view of glucose regulation in the whole animal. You should then be able to describe the hormonal control mechanisms at work in maintaining the 24 hour glucose profile that we began with in Figure 1. A number of sites that control the blood glucose level have been identified in Sections 2 and 3.

☐ What are they?

■ 1 The hypothalamus through the autonomic nervous system to the pancreas.
 2 The gut via hormones which act at the pancreas.
 3 The islets of the pancreas itself.
 4 The liver and other hormone target tissues (mainly muscle and fat cells).

The one human activity that most regularly raises the blood glucose level is eating a meal. Anticipation of the meal will stimulate insulin secretion from the pancreas (Section 3.1) via the parasympathetic system, and the presence of food in the digestive system triggers hormone release from the gut, which again stimulates insulin secretion (Section 2.1). Therefore, before glucose is actually absorbed, insulin levels are already raised in the blood. The rising level of blood glucose then triggers release of more insulin from the pancreas by acting directly within the B cells (Section 2.1). Insulin release will inhibit glucagon release from the A cells, so the production of glucose in the liver is prevented (Section 2.2).

Insulin acts relatively quickly to remove excess glucose and amino acids from the blood. Glucose transport into liver, muscle and fat cells is enhanced along with the transport of other nutrients, and insulin preserves the enzyme glycogen synthase in its active form and so promotes glycogen synthesis. At the same time, insulin deactivates glycogen phosphorylase, and this ensures that the glycogen being formed is not immediately broken down again.

18

As the blood glucose level falls, insulin secretion from the pancreas slows. Lowering the level of blood insulin releases the pancreatic A cells from inhibition, and glucagon is released. If blood glucose is required for exercise or during the fast between meals, glucose is 'pulled out' of the liver glycogen stores and at the same time gluconeogenesis from amino acids and glycerol is promoted in the liver, all through the action of glucagon. The free fatty acids released from fat breakdown contribute acetyl-CoA, which can also be metabolized in the liver. Muscle is a one-way system for glucose: it goes in, but cannot be returned to the blood. The blood glucose level is also monitored by hypothalamic glucoreceptors, which activate the appropriate pancreatic cells through the autonomic nervous system.

There is nothing mysterious about the 'normal' level of circulating glucose. This value or 'set point' varies between individuals and simply represents the balanced insulin : glucagon ratio. Raise the glucose level, and the insulin : glucagon ratio rises; lower the glucose level, and the ratio falls. Remember, though, that this regulation by balance of the pancreatic hormones can easily be overriden by the autonomic nervous system, and a new level of glucose can be set in the blood.

Glucose regulation is an excellent example of feedback control, both at the level of the pancreatic islet cells (e.g. insulin inhibits glucagon, glucagon stimulates insulin, somatostatin inhibits both) and also in the target tissues (e.g. low insulin levels provoke glycogen, fat and protein breakdown, these release glucose, fatty acids and amino acids into the blood, this stimulates insulin production). This control mechanism is summarized in Figure 15.

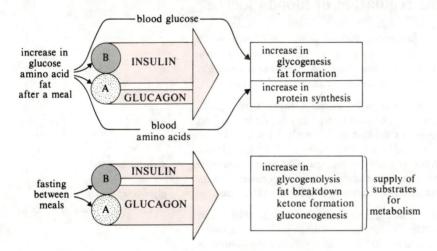

FIGURE 15   Schematic representation of the A and B cell 'couple' during extremes of glucose and other nutrients abundance and shortage, and the consequences of varying insulin–glucagon mixtures on their production and disposal.

Objectives and SAQs for Section 4

Now that you have completed this Section, you should be able to:

* identify the four main sites of control that are important in blood glucose regulation.

* describe how the levels of nutrients and pancreatic hormones in blood change from the time of anticipation of a meal until about two hours after the meal.

To test your understanding of this Section, try the following SAQs.

SAQ 5 (*Objectives 5 and 6*)   A classic physiological response to stress is a raised blood glucose level. By which mechanisms could stress raise blood glucose?

SAQ 6 (*Objectives 5 and 6*)   In an experiment to investigate stress-induced hyperglycaemia, some American research workers perfused hormones singly, or as mixtures, into the vein of a dog for 5 hours and then measured the increase in blood glucose concentration above the normal level. The results are summarized in Table 2 (overleaf).

TABLE 2 Increase in blood glucose concentration after various hormone treatments

Hormone perfused	Rise in blood glucose level mg per 100 cm^3
adrenalin ($0.05\,\mu g\,min^{-1}$)	30
glucagon ($3.5\,ng\,min^{-1}$)	10
cortisol ($4\,\mu g\,min^{-1}$)	3
glucagon + adrenalin	58
cortisol + adrenalin	58
cortisol + glucagon	35
cortisol + glucagon + adrenalin	140

Briefly compare the effects of these hormones when acting singly and when acting together. What do the data indicate about the response of body tissues to these hormone mixtures?

synergistic effect of hormones on blood glucose*

5 Diabetes—errors in the regulation of blood glucose

In this Section, the variety of causes underlying diabetes mellitus are discussed as illustrations of what happens when things go wrong in a feedback control system. The topic is expanded in the television programme Diabetes: restoring the balance, which deals with the treatment and the complications of diabetes.

The control of blood glucose levels involves an intricate network of signals, responses and feedback systems. In such a control system, any failure in one part may, therefore, be magnified throughout the system. In nearly 2 per cent of the population of the UK, blood glucose is not properly regulated—this results in the condition known as *diabetes mellitus*. Diabetes is the sign of an error in normal glucose regulation and the common symptoms of diabetes are a persistently high blood glucose level and *glycosuria*—that is, excretion of glucose in the urine.

diabetes mellitus*

glycosuria*

You will remember from Section 1 that the glucose level in human blood can vary between 80 and 110 mg per 100 cm^3 during the day. In some diabetics, blood glucose levels can rise well above 180 mg per 100 cm^3 and so produce the symptoms just described. The disorder can be fatal unless checked.

☐ From what you have now learned about blood glucose regulation, can you provide several different explanations for persistently high glucose levels? (In other words, list the various factors that influence glucose levels, and so deduce various points in the system where malfunctions could occur.)

■ This question should make you realize how intricate the control systems are for blood glucose. Our explanations fall into two groups: Those concerned with the pancreas, and those concerned with target cells. Persistently high glucose levels could be the result of:

The absence of pancreatic B cells (which normally secrete insulin); insensitivity of B cells to glucose; overproduction by pancreatic A cells (which normally secrete glucagon); overproduction by pancreatic D cells (which normally secrete somatostatin).

Lack of insulin receptors on the target cells; an inability to store glycogen; supersensitivity of receptors on the target cells to glucagon.

Any of the conditions listed above could result in persistently high levels of blood glucose. Diabetes is not, in fact, due to a single identifiable error in the control system, but is rather the result of a number of errors.

One rapid method of identifying the error in diabetes is to measure the level of insulin in the blood of diabetics after giving glucose. This has revealed two types of diabetic (see Figure 16).

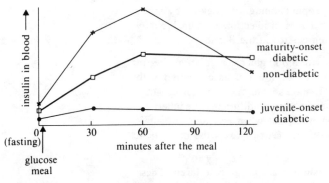

A group which has a very low level of insulin or none at all and a glucose meal therefore will not stimulate insulin secretion.

A group in which insulin levels are normal or just above or below normal and respond to a glucose meal.

The first group of diabetics commonly consist of young people who develop the symptoms before the age of 20. These are often referred to as *juvenile-onset diabetics* and will respond to insulin treatment (they are insulin-dependent). The second group develop diabetes later in life, after the age of 40, and are therefore referred to as *maturity-onset diabetics*. This group do not necessarily depend on insulin treatment (they are non-insulin-dependent) and are a rather mixed group, their diabetes being the result of one of a number of errors in glucose regulation. In the TV programme on diabetes and in other texts, the two groups of diabetics are sometimes called Type I (juvenile-onset) and Type II (maturity-onset) diabetes.

juvenile-onset (insulin-dependent) diabetes*

maturity-onset (non-insulin-dependent) diabetes*

We consider the juvenile-onset diabetics first. Their lack of an insulin response to glucose (Figure 16) appears to be due to a massive loss of pancreatic B cells (see Table 3). The lack of insulin means that glucose cannot be cleared from the blood after a meal and that glycogen breakdown will be promoted together with gluconeogenesis from the breakdown of protein and fat. The lack of insulin also means that glucagon release is not regulated, and so glucagon exaggerates the breakdown of glycogen and production of glucose.

TABLE 3 Percentage of A, B and D cells in islets of juvenile-onset diabetics and in non-diabetics

	A	B	D
non-diabetics	29	61	10
juvenile-onset diabetics	76	—	24

☐ What will happen to body fat in the absence of insulin?

■ Fat breaks down in the absence of insulin, thus releasing fatty acids, which form ketone bodies.

If ketone bodies build up in the blood, *ketosis* may result, which will lead to *coma* due to the increased acidity (low pH) and the loss of body fluid associated with glycosuria. Happily, the insulin-dependent diabetic's condition can be controlled by *insulin treatment*.

ketotic coma*

insulin treatment of diabetes*

☐ How would you give insulin to a diabetic?

■ Insulin is a peptide, and therefore would be broken down in the gut if given orally. Insulin is usually administered by injection into muscle, under the skin or sometimes directly into a vein.

The discovery of insulin in 1921 dramatically reduced death from diabetes in these insulin-dependent cases. Most commercial insulin is prepared from beef or pig pancreas, and insulin of high purity and improved long acting preparations are now available.

☐ Why should the purity of insulin matter?

■ Impurities (e.g. other proteins) may produce an allergic reaction in the body.

The cause of the loss of B cells in young people is being investigated by many research groups. There is growing evidence that viral infection early in life may have initiated an antibody reaction that then destroyed the B cells.

Maturity-onset diabetics develop problems with blood glucose regulation later in life. Their insulin levels can be normal, raised or lowered as compared with non-diabetics. These older diabetics do not develop ketosis and do not necessarily require insulin injections. This maturity-onset diabetic condition may be inherited. A study of twins carried out in London showed that when one twin of a pair developed diabetes after the age of 50, in 90 per cent of cases the other twin developed diabetes within a few years.

Some maturity-onset diabetics have high insulin *and* high glucose levels. These diabetics are often obese. The problem in these cases seems to be that excess glucose intake in the diet produces a persistently high level of insulin secretion from the pancreas.

☐ What effect would persistently high levels of insulin have on liver cells?

■ Remember from Section 2 that insulin receptors become reduced in number in response to a persistently high level of insulin in the blood.

By controlling the dietary intake of carbohydrate and so reversing the obesity, insulin secretion from the pancreas is reduced, and the target cells such as the liver again become sensitive to insulin.

In maturity-onset diabetics who secrete low amounts of insulin, the problem may be a lack of response by B cells to glucose, or a decreased ability of B cells to synthesize and secrete insulin. Drugs, such as sulphonylureas, can be used to stimulate insulin secretion.

Summary of Section 5

Diabetes is a condition where blood glucose regulation has failed at any one of a number of control sites. Two broad types of diabetic can be identified and the differences are summarized in Table 4.

TABLE 4   Differences between juvenile-onset and maturity-onset diabetics

Features	Juvenile-onset	Maturity-onset
age at onset (years)	under 20	over 40
proportion of all diabetics diagnosed	less than 10 per cent	more than 90 per cent
appearance of symptoms	acute	slow
ketosis	frequent	rare
obesity at onset	uncommon	common
B cells	few	variable
insulin	low blood level	variable
family history of diabetes	uncommon	common
antibody to islet B cells	present in blood	absent

Objectives and SAQs for Section 5

Now that you have completed this Section, you should be able to:

★ distinguish between juvenile-onset and maturity-onset diabetes.

★ explain why insulin treatment would be inappropriate for some forms of diabetes.

To test your understanding of this Section, try the following SAQs.

SAQ 7 (*Objective 7*)   In juvenile-onset diabetics, glucagon levels are higher than in non-diabetics. (a) What would the effect of this raised glucagon be on blood glucose level? (b) Suggest two hormone treatments that could reduce blood glucagon levels.

SAQ 8 (*Objectives 2, 4 and 7*)   Outline the events that result in ketotic coma in juvenile-onset diabetes. (To answer this question, you will need to draw on information in Sections 1.2 and 2.1 as well as Section 5.)

SAQ 9 (*Objective 7*)   What complications can you think of that may be associated with insulin therapy in diabetes? (Think carefully about what type of insulin is given and the general effects of insulin on blood glucose.)

6   Concluding remarks

In this Unit, we have looked in some detail at the mechanisms responsible for maintaining a steady level of glucose in the blood, and at the problems that arise when homeostatic regulation breaks down. Glucose, of course, is not the only substance whose level is so finely controlled. We could have chosen any one of a number of ions, for example, and later Units in this Course will discuss the role of the kidney in the maintenance of ion and water balance in the body. Similarly, the levels of oxygen and carbon dioxide in the blood are finely regulated to meet tissue demand with an adequate supply.

Glucose regulation is achieved by the concerted efforts of the endocrine and nervous systems and a complex series of feedback mechanisms are in operation at the level of the pancreatic cell, in the liver and in the brain. Homeostasis is the maintenance of a particular level or set point. New levels can be set at any time to cope with increased demand.

There are other physiological processes controlled and coordinated by the endocrine and nervous systems that do not have distinct and constant set points. Many physiological processes are 'primed' by the action of hormones at a particular time (e.g. the determination of phenotypic sex in mammals). Others are composed of a sequential series of responses until a conclusion is reached—in other words a cycle (e.g. the menstrual cycle). The way that the endocrine and nervous systems achieve control of the mammalian reproductive cycle is an extremely clear example of how these two systems are integrated, and this topic forms the basis of the next Unit.

Objectives of Unit 17

Now that you have studied this Unit, you should be able to:

1   Define, recognize or place in the correct context the terms marked with an asterisk in the Table A.

2   Describe the effects of insulin, glucagon, corticosteroids, adrenalin and glucose on the interconversion of glucose and glycogen. (*SAQs 1, 4 and 8*)

3   Explain, in words or with a flow diagram, how glucagon, insulin and somatostatin influence each other's secretion. (*SAQs 2 and 3*)

4   Describe the effects of glucagon and insulin on protein and fat stores. (*SAQs 3, 4 and 8*)

5   Describe the mechanisms whereby stress or trauma can raise blood glucose levels. (*SAQs 5 and 6*)

6   List the four main sites of control in blood glucose regulation. (*SAQs 5 and 6*)

7   Distinguish between the two main types of diabetes in terms of their underlying cause. (*SAQs 7, 8 and 9*)

SAQ answers and comments

SAQ 1   The accurate statements are (iii) and (iv).

(i)   Glycogen phosphorylase and synthase do exist as *a* and *b* forms, but the activation of synthase depends on the *inhibition* of the cyclic AMP dependent protein kinase.

(ii)   Glycerol and some amino acids are substrates for gluconeogenesis by reversed glycolysis, but fatty acids are not.

(v)   As the concentration of glucose rises in liver cells, the level of phosphorylase *a* drops and synthase *a* rises.

SAQ 2   (a), (c) and (d) would all promote insulin secretion; (b) and (e) would prevent insulin secretion.

(a)   Somatostatin inhibits insulin secretion by a direct effect on the B cell; removing the inhibition stimulates release.

(b)   Glucose stimulates insulin secretion, but the effect is via a regulatory protein that is dependent on Ca^{2+} activation of calmodulin.

(c)   Cyclic AMP promotes insulin production, and inhibiting phosphodiesterase (the enzyme that breaks down cyclic AMP) would result in insulin secretion.

(d)   The presence of food in the digestive system stimulates the release of gut hormones, which in turn promote insulin secretion.

(e)   Glucose must enter cells to activate Ca^{2+}-dependent insulin secretion.

SAQ 3   (i) and (ii) would result in fat and protein breakdown.

(i)   The absence of B cells implies an insulin deficiency, which means that fat and protein are broken down for use in the liver.

(ii)   Glucagon promotes the breakdown of fat and protein for gluconeogenesis.

(iii)   A carbohydrate-rich meal results in a plentiful supply of glucose and an *increase* in fat stores if anything.

(iv)   Somatostatin normally inhibits insulin and glucagon secretion, and removing the inhibition would result in a rise in insulin and so the protection of fat and protein stores.

SAQ 4   (a)   Insulin, by activation of glycogen synthase; glucocorticoids, by deactivation of glycogen phosphorylase.

(b)   Glucagon, by activation of glycogen phosphorylase and also by activation of enzymes in the TCA cycle; adrenalin, by activation of glycogen phosphorylase.

(c)   Glucocorticoids and glucagon.

SAQ 5   Stress involves a rise in levels of glucocorticoids and of adrenalin. Glucocorticoid release from the adrenal cortex is stimulated by the action of CRF (from the hypothalamus) on the pituitary (which produces ACTH). Glucocorticoids produce a small rise in blood glucose. Stress via the sympathetic nervous system stimulates the adrenal medulla to release adrenalin, which promotes liver gluconeogenesis; also the sympathetic system inhibits insulin production directly in the pancreatic islets and stimulates glucagon release.

SAQ 6   When injected singly, the three hormones each raise blood glucose concentration. In combinations, any two together raise blood glucose by a greater amount than the sum of their individual effects (e.g. adrenalin 30 and cortisol 3, but adrenalin + cortisol 58). When all three are injected together, the rise in blood sugar is between two and four times greater than their individual effects or their combined effects as pairs. These are examples of the *synergistic* effects of hormones (Unit 16, Section 8.2.2).

SAQ 7   The glucagon level in juvenile-onset diabetes is due to the absence of insulin, which normally inhibits glucagon release. Therefore administration of insulin would depress glucagon levels. Somatostatin normally inhibits glucagon secretion, so by giving somatostatin, we would expect to depress glucagon levels.

SAQ 8   1   In the absence of insulin, protein and fat is broken down to amino acids, fatty acids and glycerol, which are metabolized mainly in the liver.
2   In the absence of glucose metabolism, the liver cannot convert all the acetyl-CoA produced from fatty acids via the TCA cycle.
3   Acetyl-CoA in excess will produce ketone bodies such as acetone, β-hydroxybutyrate and acetoacetate.
4   These ketone bodies (which are strong organic acids) enter the blood and cause changes in blood pH.
5   Excess glucose in the blood as a result of insulin deficiency causes glycosuria and water is lost from the body through the kidneys. The kidneys also attempt to excrete excess ketone and therefore increase this loss of water.
6   Massive dehydration caused by the loss of water (and salts) in the kidneys leads to osmotic problems for cells, particularly in the brain, and coma can result.

SAQ 9   Insulin will lower the level of blood glucose. If too much is given, then *hypoglycaemia* may result, leading eventually to coma. If too little insulin is given to maintain blood glucose within the correct limits, then after a meal, glucose levels will rise towards the hyperglycaemic state, and there is the danger of *ketotic coma*. Diabetics need to balance their insulin requirements with their diet and physical activity very carefully. There is yet another problem with insulin therapy mentioned in Section 5. Commercial insulins are not of human origin, but are purified from pig or beef pancreas. Although very similar to human insulin, commercial insulins given over long periods may initiate the formation of antibodies to the insulin. These antibodies will tend to deactivate subsequent insulin injections. Happily, the purity of insulins is now such that although antibodies are formed, they do so to a small extent, and insulin insensitivity is not a major problem in insulin-dependent diabetes.

Further reading

These books are relatively inexpensive and the level at which they treat the topic should now be well within your reach:

R. N. Hardy (1981) *Homeostasis* 2nd edition. Studies in biology, Edward Arnold, London.

A. L. Notkins (1979) The Cause of Diabetes, *Scientific American*, vol. 241 (no. 5), p. 56.

W. G. Oakley, D. A. Pyke, and K. W. Taylor (1978), *Diabetes and its Management*, 3rd edition, Blackwell, Oxford.

M. I. Drury (1979) *Diabetes Mellitus*, Blackwell, Oxford.

Acknowledgements

Grateful acknowledgement is made to the following for permission to reproduce material in this Unit.
Figure 2   Current Topics in Cellular Regulation, B. L. Horecker and E. R. Stadtman (Eds), Academic Press, 1976; *Figure 4*   J. F. Lamb et al., *Essentials of Physiology*, Blackwell, 1980;

Figures 5, 12 and 15   Glucagon and the A cell, R. H. Unger et al., in *Recent Progress in Hormone Research* (R. O. Greep, Ed.), Academic Press, 1973; *Figure 13*   Insulin, Glucagon, and Somatostatin Secretion in the Regulation of Metabolism, R. H. Unger and R. E. Dobbs, in *Annual Review of Physiology* (I. S. Edelman, Ed.), Annual Reviews Inc., 1980.

unit 18

Control Mechanisms in Reproduction

Contents

TABLE A   Scientific terms and principles used in Unit 18

Assumed knowledge	Introduced in an earlier Unit	Unit	Introduced or developed in this Unit	Page
abortion	ablation	16	accessory glands (prostate, Cowper's, seminal vesicles)*	8
auditory stimuli	adrenal gland	16	amenorrhoea	29
birth control	adrenocorticotropic hormone		androgen*	10
castration	ACTH	16	androgenital syndrome	11
configuration (D and L)	agonist	16	androstenedione	10
contraception	amines	16	breeding cycle*	6, 9
copulation	amino acid	5	breeding season*	6
coitus	analogues	16	Bruce effect*	24
egg	androgen	16	breeding life span	6
ejaculate	antagonist	16	cervix (pl. cervices)	6
embryo	antibodies	5	combined-type pill*	28
epithelium	antigens	5	continuous breeders*	8
erectile tissue	blastocyst	11	contraception*	27
fecundity	cholesterol	16	corpus luteum (pl. corpora lutea)*	6
fetus, fetal	chromosomes (X, Y)	11	day length*	20
fertility drug	corticosteroids	16	delayed implantation*	6, 22
formulation of drugs	corticosterone	16	dihydrotestosterone	25
genes	cortisol	16	dopamine*	17
genetics	cyclic AMP	16	endometrium*	6
genitalia	diploid	11	endopeptidase	29
gonads	differentiation	11	environmental cues*	19
hysterectomy	dopamine	16	epididymis	8
latitude	feedback system	7, 16	Fallopian tube (oviduct)*	4
litter of young	fertilization	11	female reproductive system*	4
mammary glands	follicle stimulating hormone (FSH)	16	fertility drugs*	14
mating			fimbria (funnel) of Fallopian tube	5
meiosis	gametes	11	follicular phase of cycle*	13
menstrual cycle	glycoprotein	5	follicle-stimulating hormone (FSH)*	10
menopause	gonadotropin	16	follicle stimulating hormone/ luteinizing hormone releasing factor (FSH/LH-RF)*	11
milk	gonadotropin releasing factor (FSH/LH-RF)	16		
nipple	growth hormone GH	16	gestation period*	7
olfactory stimuli	hibernation	3	Graafian follicles*	5
ovary	hypothalamus	16	granulosa cells	5
parental care	haploid	11	gonadotropin*	10
penis	immunological reactions	5	hypophysectomy*	10
placebo	luteinizing hormone (LH)	16	hypothalamic nuclei*	11
placenta	mammals	3	hypothalamus*	11
plasma (blood)	marsupials	3	implantation (of blastocyst)*	6
pregnancy	messenger molecule	16	induced ovulators*	22
puberty	negative feedback	16	interstitial tissue, cells	5
rhesus factor	neurosecretion	16	Leydig cells*	8
scrotum	neuron	16	luteinizing hormone (LH)*	10
sperm, spermatozoa	neurotransmitters	16	long-day breeders*	20
sterility	noradrenalin	16	luteal phase of cycle*	13
suckling	oestrogens	16	luteolysis*	6
tactile stimuli	oocyte	11	luteolytic factor	14
the (contraceptive) pill	oogenesis	11	luteotropic factor	14
uterus	ovum (pl. ova)	11	male reproductive system*	8
vagina (pl. vaginae)	pituitary	16	masculinization*	26
	positive feedback	16	melatonin*	21
	primates	3	menstrual cycle*	6
	primary oocyte	11	mini-pill*	28
	progestin	16	monoestrus	6
	prohormone	16		
	prostaglandin	16		

* These terms must be thoroughly understood—see Objective 1.

Study guide for Unit 18

Unit 18 completes the series of three Units (16, 17 and 18) concerned with the regulation and control of physiological processes; it represents one week of study. While regulation and control is the principal theme, you will find structure–function relationships emphasized, particularly in Section 1, which deals with the reproductive tract.

The physiology of reproduction draws on information from a number of disciplines, and to incorporate these, the Unit concentrates principally on the control of the production and release of sperm and eggs and how these events are 'timed' both in the short term, to improve the chances of fertilization, and seasonally, to improve the chances of survival of the offspring. It would be useful for you to revise Unit 11, Section 4.1, where the major events in gametogenesis and development are outlined. The Unit is limited to reproduction in mammals.

Section 1 is short (about 1 hour of study time), and it describes the anatomy of the reproductive tract and introduces the various types of breeding cycles and breeding seasons shown by different mammals. Section 2 is the most important part of the Unit and you should allow 2½ hours for its study. It demonstrates the involvement of communication systems (covered in Unit 16) in gamete production and the maintenance of the zygote after fertilisation. A lot of the hormones and concepts were covered in Unit 16, and you are referred back to this Unit (particularly Sections 7 and 8) a number of times in Section 2. Sections 3 and 4 are much shorter (less than an hour's study each), and they discuss the involvement of nerves and hormones in breeding cycles, particularly where reproductive status closely follows seasonal changes in light, and in sexual determination and maturation. The final Section looks at how steroids and peptides can act as contraceptives; its study again should be less than an hour.

Mammals have been studied more extensively than most other groups of animals (with the possible exception of birds) for clinical and economic reasons. While mammals show many similarities in reproductive physiology and the underlying control mechanisms, it is not sensible to search for a unified scheme that would fit all mammals. The group demonstrates great diversity in the control mechanisms (and the environmental cues) that regulate reproduction, and it is often these species differences that add to our understanding of reproductive physiology. You should, therefore, beware of extrapolating information given here to all mammals. In most Sections of the Unit, the emphasis is on the human or the rat, but descriptions of other species are provided to demonstrate this variety.

The television programme associated with this Unit, *Only in the Mating Season*, is concerned with the environmental cues to which seasonal breeders are sensitive and shows how light can stimulate endocrine and neural changes that bring about synchronization of reproductive readiness in individuals. The close links between hormones and the nervous system are emphasized in the programme, as they are in the three Units (16–18) as a whole.

There is a Home Experiment associated with this Unit, and it deals with the effect of day length (photoperiod) on the reproductive system of the female ferret. The experiment involves using the microscope in the Home Experiment Kit and the relevant prepared histological slides. It will take 2–3 hours, and it is best to tackle it when you have finished the Unit.

1 Systems and cycles

This Section provides basic information about the reproductive organs of mammals and the cycles of events in sexually mature males and females. If you already know about these, you could omit the Section, but answer the SAQs to make sure you have achieved the objectives. Later sections assume knowledge of the information given here.

A number of factors contribute towards reproductive success. These include mechanisms that determine successful pregnancy and that maximize fecundity. Parental care is a common feature in mammals, and it takes the form of rearing the young in a sheltered environment and providing them with food.

Reproductive success also requires timing, not only of gamete production, but also of the behaviour of sexually mature individuals within a species. Communication of the reproductive condition of one individual to another may involve complex behavioural displays and mass movements of groups of animals from one place to another. Complex endocrine changes underlie all these events, and the relationships between hormones, the nervous system and behaviour re-emphasizes the theme we traced through Unit 16.

Before looking in detail at the mechanisms involved in controlling reproduction, we introduce the major target organs—the gonads—that produce the gametes. In all mammals, there are two distinct sexes.

1.1 Female mammals

Figure 1 shows the *female reproductive system*. It consists of paired *ovaries* and a series of ducts that receive the eggs released from the ovary and convey them to the site of implantation, the uterus (or womb). The *Fallopian tubes* (*oviducts*) also receive sperm and convey it to the site of fertilization within the oviduct itself.

female reproductive system*
ovary*
Fallopian tube (oviduct)*

In all mammals, the ovaries are paired, and their size depends largely on the age and 'reproductive state' of the female. Within the ovaries there are *primordial follicles* (see Figure 2) containing the cells that give rise to the egg and its surrounding membranes. These follicles are present in very large numbers early in development—more than can ever ripen and be shed. Primordial follicles ripen into

primordial follicles*

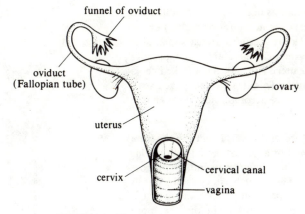

FIGURE 1   The female reproductive tract.

large *Graafian follicles* and burst to release the egg at the time of ovulation. Between the follicles lie spindle-shaped *interstitial cells*, which produce steroid hormones. The primordial follicles each contain a primary oocyte surrounded by a single layer of follicular or *granulosa cells*. As the follicle matures and the oocyte enlarges, the surrounding follicular cells enlarge and multiply. The *stromal tissue* around the follicle differentiates into a vascular inner layer made up of secretory cells and an outer layer of connective tissue. These two layers make up the *theca*. Both the secretory cells of the theca and the follicular cells produce *steroid hormones*.

Graafian follicles*
interstitial tissue, cells

granulosa cells
stromal tissue

theca of follicle
steroid hormones

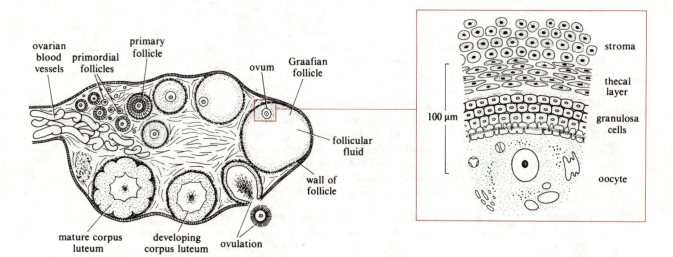

FIGURE 2   The production of follicles in the ovary. The cycle of production, ovulation and corpus luteum formation is shown clockwise around the ovary. Inset shows the wall of a mature follicle.

In most primates, usually only one follicle ripens at a time, though occasionally several may ripen, shedding a number of eggs together at *ovulation*. Many mammals (e.g. dog, cat, pig, ferret, rabbit) regularly produce a number of eggs from each ovary at each ovulation and the elephant shrew sheds over 100 eggs at one time! In the nine-banded armadillo, a single egg is shed, but after fertilization the resulting zygote divides into four.

☐ How would the offspring from four eggs be related to one another as opposed to offspring from the division of one egg into four?

■ All four offspring of the armadillo are genetically *identical*. In the case of four distinct eggs, each would be genetically different and four dissimilar individuals would be produced.

After ovulation, the egg(s) pass through the *funnel (fimbria) of the Fallopian tube*. Fimbriae either envelop the ovaries, or are close to them (as in Figure 1) and show some mobility at the time of ovulation. Fimbriae can pick up eggs lost in the body cavity or ovulated from the opposite ovary. The wall of the oviduct is lined with ciliated epithelium that beats away from the ovary, creating a 'current' towards the uterus. The walls of the oviduct are muscular, and perform a rippling motion of contraction and relaxation that helps propel the eggs down the oviduct. After fertilization, the egg divides to form a blastocyst.

fimbria (funnel) of Fallopian tube

The oviducts lead into a *uterus*, which may be single as in primates or double as in rats and mice (see Figure 3). The *cervix* (or cervices in the rat or mouse) opens into the vagina, which opens externally, guarded by the *vulva*. In marsupials, for example the opossums of North and South America, there are two externally opening vaginae, two cervices and two separate uteri. The male has a forked penis capable of entering the two vaginae simultaneously. This arrangement gave rise to a superstition that copulation in the opossum is accomplished through the nostrils!

Whether or not a fertilized egg implants in the uterus depends largely on the state of the epithelium that lines it—the *endometrium*. For a limited period after ovulation, the endometrium is highly receptive to anything lying on it, and it will grow around the blastocyst. It is not clear why only a few of the blastocysts implant in the species that shed many eggs at a time. In some species (e.g. rabbit and pig), it is common to find some early embryos being reabsorbed by the mother, i.e. more blastocysts implant than are carried through to term (birth). Some species show *delayed implantation*, in which the fertilized blastocyst remains unattached for varying periods. The adaptive significance of delayed implantation will become clear in Section 3.

In women, ovulation takes place about every 28 days, from the age of 12–14 (usually) until menopause (45–50). The monthly or *menstrual cycle* is so called because the bleeding (menstruation) at the start of the cycle is a monthly occurrence. (Menstruation can be considered the end or the start of a cycle. In this Unit, menstruation is designated day 1 of a new cycle.) In mammals, the cycle is usually called the *oestrous cycle*, because many mammals show *oestrus* or 'heat' (a measurable rise in body temperature) and characteristic behaviour at about the time of ovulation.

Follicular development in the human cycle takes 12–15 days. From about day 4 (day 1 being the first day of menstruation), the endometrium begins to thicken and becomes vascularized (supplied with blood vessels). After ovulation (usually near the mid-point of the cycle), further build-up of the endometrium takes place (particularly an increase in vascularization), and in the ovary, the ruptured (empty) follicle develops into a solid secretory body called the *corpus luteum* (see Figure 2). If the egg is fertilized and *implantation* occurs in the uterus, the corpus luteum in the ovary persists. In some species, the corpus luteum persists throughout pregnancy, while in others, it remains only through the early stages of pregnancy. The corpus luteum will also persist if the animal is *pseudopregnant*—that is, behaving physiologically as if pregnant when actually it is not. Pseudopregnancy is common in the rat, rabbit and ferret, as we see later in Section 2.3.

If pregnancy does not occur, the corpus luteum breaks down—a process called *luteolysis*. In the human cycle, this occurs at about day 26 or 27. The uterine endometrium breaks down, shedding into the uterine lumen blood, epithelium, capillaries and other tissue, most of which is then voided as the menstrual flow.

The length of oestrous cycles and their frequency or 'seasonality' depends, as does the length of breeding life, on the species. There are, in effect, three periods in reproduction: a *period of breeding potential* in the lifetime of the animal, within that period possibly a *breeding season* or a *seasonal cycle* (although many animals may breed continuously), and within that season one or a series of oestrous cycles. The cycle itself can be a single one (*monoestrous*), or there can be two, or many (*polyoestrous*). The duration of a cycle also varies between species.

When implantation and pregnancy occur, the corpus luteum persists and the uterine endometrium continues to develop. A complex *placenta* (from the Greek for 'flat cake') is formed as a result of intimate apposition of fetal tissues to maternal tissues (see Figure 4). The placenta functions as an exchange network for fetal and maternal blood, but the two do not actually mix.

☐ Why would it be disadvantageous to allow the two blood supplies to mix?

■ There could be immunological reactions. The mother and fetus are not genetically *identical*, and therefore the mother may develop antibodies against the fetus and thus reject it. This can happen in some species (including ourselves) where this 'barrier' is not complete, and a developing fetus with blood antigens distinct from the mother (e.g. the 'rhesus' factor) inherited from the father may be destroyed.

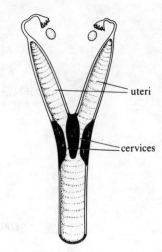

FIGURE 3 The uterus of a rat or mouse; note the paired cervices and uteri.

menstrual cycle*

oestrous cycle*
oestrus*

corpus luteum (*pl.* corpora letea)*
implantation (of blastocyst)*

pseudopregnancy*

luteolysis*

breeding life span
seasonal cycle*
breeding season*
monoestrus
polyoestrus

placenta

6

Respiratory gases are exchanged across the placenta together with nutrients and nitrogenous waste. The placenta is also an important source of hormones, particularly in those mammals where the corpus luteum does not necessarily persist throughout pregnancy. The precise form that the placenta takes depends on the species.

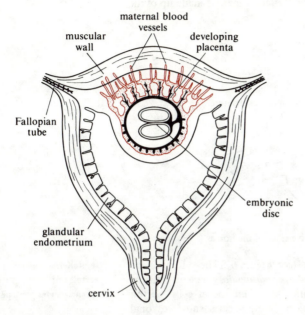

FIGURE 4 A section through the human uterus to show the position of implantation and the placenta.

The *gestation period* (duration of pregnancy) also varies between species and depends on two factors: the size to which the fetus grows and the stage of development it reaches before *parturition* (birth). In the rat and hamster, animals of roughly similar size, the gestation periods are 21 and 18 days, but in the guinea pig, it lasts 68 days. The explanation for the difference lies in the stage of development reached by the fetus at birth. Rats and hamsters have small, naked young with closed eyes and ears, and they are barely mobile. Guinea pig young are born like small adults. The gestation period is remarkably consistent within a species, varying by only 5–10 per cent. The timing of birth is intimately controlled by both maternal and fetal *neuroendocrine systems*.

gestation period*

parturition

neuroendocrine systems

Table 1 illustrates the variation in oestrous cycle, gestation period and number of young in mammals.

TABLE 1   The length of the oestrous cycle and the gestation period (days) and the number of offspring for several mammals

Species	Oestrous cycle	Gestation period	Offspring
European hedgehog	— (monoestrous)	34–49	5
Rhesus monkey	28 ⎱ (polyoestrous—	164	1–2
chimpanzee	34 ⎰ continuous)	237	1–2
woman	28 ⎰	280	1–2
Indian elephant	(polyoestrous)	623	1–2
blue whale	—	365	1–2
domestic pig	21 (polyoestrous— continuous)	113	6–11
domestic cattle	21 (polyoestrous— almost continuous)	282	1–2
domestic sheep	17 (polyoestrous— seasonal)	150	1–3
ferret	*none	42	9
domestic cat	*15–21	63	4
domestic rabbit	*none	31	6
laboratory rat	4–6	21	7–9
golden hamster	4	18	6
guinea pig	16	68	3
laboratory mouse	5	19	6

* These are induced ovulators—see Section 3

1.2 Male mammals

The *male reproductive system* consists of a pair of *testes*, paired *accessory glands* and a duct system which includes the penis (Figure 5). During development, the testes descend into the scrotum, and during adult life, they function as producers of sperm and hormones. Some 90 per cent of the testicular mass is made up of the

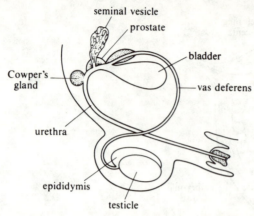

FIGURE 5   The male reproductive tract of the horse.

seminiferous tubules that produce the gametes (see Figure 6). These tubules are very convoluted and comprise *Sertoli cells* and *spermatogonia*, which give rise to *spermatocytes*, which in turn become *spermatids* and then sperm cells. Sertoli cells aid in the process of sperm cell formation. Many species show seasonal spermatogenesis; at other times of the year, the testes are inactive. Between the tubules are connective tissues and the *Leydig cells*; these secrete steroid hormones and are seen in all male mammals in breeding condition.

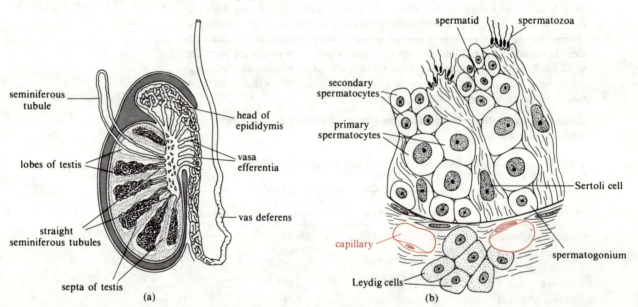

FIGURE 6   (a) A section through the testis. (b) At high power, part of a seminiferous tubule and the interstitial zone.

In all mammals (except marine species and the elephant and rhinoceros), the testes descend into the scrotum. In *continuously breeding* species, they remain in the scrotum, whereas in *seasonally breeding* species, they ascend into the body cavity during the non-breeding season. The main function of the scrotum appears to be to provide the testes with an environment somewhat cooler (by 2–3 °C) than the body cavity. Additionally, the arrangement of arteries and veins to the testes functions to maintain a cooler temperature. Shepherds learned long ago that it was possible to cause temporary sterility in rams by physically tying the testes close to the body wall. Similarly, if the testes do not descend, then that individual is often sterile. This effect of temperature on the testes appears to be due to the inability of the Leydig cells to secrete hormones.

The seminiferous tubules fuse into a single tube, the *epididymis*, which becomes a muscular *vas deferens*. The two vasa deferentia empty into the *urethra*, a tube common to the urinary and genital systems, which receives secretions from the accessory glands (prostate, Cowper's glands and seminal vesicles). The urethra runs through the penis, which is erectile. These accessory glands develop to different degrees in different mammals. No seminal vesicles are present in the cat, dog and wolf, but the prostate is well developed. There is considerable variation in the amount of seminal fluid contributed by the various accessory glands and testes, but in any species not more than 2–5 per cent of the total ejaculate is made up of the secretion of the testes and epididymis. In man, the prostate contributes 15–30 per cent and the seminal vesicles 40–80 per cent of the total ejaculate. *Seminal fluid* has two functions: it serves as a suspending and activating medium for the sperm cells, which are not motile up to this time, and it provides the cells with a medium rich in ions, vitamins and the sugar, fructose.

seminal fluid

Male mammals, like females, are either active continuously or only during a *breeding season*. The behaviour of male rabbits suggests that, within a breeding season, there is a cycle or rhythm of testicular function, although this rhythm (if it exists) is neither as accurately timed, nor as well defined, as the oestrous cycle of the female.

breeding cycle*

Summary of Section 1

Ripe eggs are released from the ovary and pass down the Fallopian tube (where fertilization can take place) to the vascularized uterus. Fertilized eggs will implant and a placenta is then formed. If fertilization does not occur, the wall of the uterus breaks down and in the human is shed as the menstrual flow. The length of an oestrous cycle varies between species, many of which show seasonal cycles; the length of gestation also varies between species.

Sperm are produced in the testes continuously or during a breeding season and accessory glands add secretions to the seminal fluid as this passes from the vas deferens into the urethra. The Leydig cells in the testes produce steroid hormones, and the Sertoli cells aid in the process of sperm formation.

Objectives and SAQs for Section 1

Now that you have completed this Section, you should be able to:

★ describe, in words and diagrams, the basic anatomy of the male and female reproductive systems.

★ distinguish between the terms oestrous (menstrual) cycle, breeding season and gestation period.

★ list the sequence of events in a menstrual cycle.

To test your understanding of this Section, try the following SAQs.

SAQ 1 (*Objectives 2 and 3*)   Select three accurate statements from (i)–(vii).

(i) At ovulation, follicles are released from the ovary and pass down the Fallopian tube, which is connected to the ovary and the uterus.

(ii) The corpus luteum located in the uterus persists there if pregnancy occurs and aids in the maintenance of the endometrium.

(iii) The gestational period describes the time taken for a follicle to mature and ovulate.

(iv) In addition to producing sperm, the testes secrete steroid hormones from the Leydig cells.

(v) Sperm are carried to the vas deferens and urethra, which also receives secretions from accessory glands.

(vi) All male and a majority of female mammals are continuous, rather than seasonal, breeders.

(vii) Follicular development (from a primordial to a Graafian follicle) takes about 14 days in humans.

SAQ 2 (*Objective 2*)   It is not uncommon to find the Fallopian tubes partially blocked by tissue growth, which prevents the ovulated egg from reaching the uterus. Does this mean that fertilization is impossible? What complications could arise in someone with blocked Fallopian tubes?

2 The control of the reproductive cycle

This Section describes the role of the endocrine and nervous systems in the oestrous cycle. Many of the hormones involved, and their different modes of action, have been introduced in Unit 16, Sections 6, 7 and 8.

In the previous Section, we described the cyclical nature of gamete maturation and release. During the breeding season, if fertilization is successful, the cycle of ovulation is replaced by a gestation period. If the released egg is not fertilized, the ovulatory cycle is repeated. The cyclical nature of the whole process of reproduction suggests that an intricate network of control systems are involved. In this Section, the control of the oestrous (in particular, the menstrual) cycle is considered; in Sections 3 and 4, we look at the role of nerves and hormones in seasonal breeders and in the process of sex determination and maturation.

2.1 Releasing factors, stimulating hormones and steroids

If the pituitary of a sexually immature rat is surgically removed (an operation called *hypophysectomy*), then among other things, the gonads fail to become functional. If an identical operation is carried out on an adult rat, the gonads shrink and eventually atrophy. In both cases, when extracts of pituitary are injected into the hypophysectomized rats, gonadal function is restored. These early observations suggested a link between the pituitary and the gonads, and in 1931, it was demonstrated that the pituitary secretes two *gonadotropins* (Unit 16, Section 7), which were later called *luteinizing hormone (LH)* and *follicle-stimulating hormone (FSH)*.

hypophysectomy*

gonadotropin*
luteinizing hormone (LH)*
follicle-stimulating hormone (FSH)*

Pituitaries from both males and females contain LH and FSH. These are glycoproteins with a relative molecular mass of around 30 000 and both consist of α and β subunits. FSH and LH are secreted from cells in the anterior pituitary.

☐ What other tropic hormones are released from the anterior pituitary?

■ Adrenocorticotropic hormone (ACTH), thyroid stimulating hormone (TSH) and growth hormone (GH). (See Unit 16, Section 7).

LH and FSH are secreted into the blood and stimulate the gonads to produce other hormones. The ovaries and testes release steroid hormones (Unit 16, Section 6), and these have a variety of target organs and, therefore, a variety of effects.

☐ What is the parent molecule from which steroids are synthesized?

■ Cholesterol (Unit 16, Section 6.1).

☐ What are the two main groups of steroid hormones?

■ Sex steroids (oestrogens, androgens and progestins) and corticosteroids (e.g. cortisol and corticosterone).

The route of sex steroid production from cholesterol proceeds through *pregnenolone*, which is oxidized to form *progesterone*. Conversion of either pregnenolone or progesterone will produce the *androgens* (e.g. in the testes—*testosterone* and *androstenedione*) and the oestrogens (e.g. in the ovary—*oestradiol* and *oestrone*). Androgens can be converted into *oestrogens*.

pregnenolone
progesterone*
androgen*   testosterone*
androstenedione   oestradiol   oestrone
oestrogen*

☐ What significance could this link between different sex steroid hormones have in the ovary?

■ It means that testosterone may be present in the ovary as a precursor of oestrogens.

Similarly, small quantities of oestrogens can be found in the testis. While the release of oestrogens in the human male is quite small, both the boar and the stallion produce quite large quantities of oestrogens.

☐ In what other endocrine tissue might you expect to find sex steroids? (What other endocrine gland secretes steroids?)

■ The cortex of the adrenal glands has steroid-containing cells (mainly producing corticosteroids), and these also produce traces of both oestrogens and androgens.

The adrenal gland and ovary get cholesterol from the blood. The testes, however, seem to manufacture their own cholesterol (from acetate). The reasons for this difference are obscure, though it is possible that the testes are unable to gain access to the cholesterol in the blood. This blood barrier may also protect sperm in the testis from immunological attack.

Occasionally, a defect in steroid production in the adrenals leads to the production of large amounts of androgen. There is feedback from the level of androgens circulating in the blood to the pituitary, and this stops the release of gonadotropins.

□ What sort of feedback system is this?

■ A *negative* feedback system.

As the level of gonadotropins in the blood drops, so does the release of sex steroids from the gonads. Production of large amounts of androgen from the adrenal gland in a female can be disastrous.

□ Why does the production of adrenal androgen not stop along with gonadal steroid when gonadotropin levels drop?

■ The adrenals are *not* under gonadotropic control but are regulated by ACTH from the pituitary.

The negative feedback system between the adrenals and the pituitary therefore breaks down because androgens are being produced instead of corticosteroids. Now the system goes wildly out of control, because the low level of corticosteroids in the blood stimulates ACTH release from the pituitary, and more ACTH causes more androgen production in the adrenals.

□ How could this situation be quickly brought under control?

■ The problem is the *lack* of corticosteroids. If corticosteroids are given, this will lower ACTH production (by negative feedback) from the pituitary, will prevent the release of more androgen from the adrenals, and will supply the patient with the missing corticosteroids.

The result will be that gonadotropin release from the pituitary will be resumed and normal ovarian function will be restored. The condition of producing large amounts of androgen in the adrenal glands of a female, is known as the *androgenital syndrome*, and it demonstrates how catastrophic results can occur when a feedback mechanism goes slightly wrong.

androgenital syndrome

Gonadotropins from the anterior pituitary therefore regulate steroid hormone production in the gonads, but this is only one part of the control system.

□ What controls gonadotropin release?

■ The *hypothalamus* produces releasing (and release inhibiting) factors that are carried to the pituitary in the blood supply (Unit 16, Section 7.1).

hypothalamus*

□ What is the releasing factor concerned with pituitary gonadotropin release?

■ *Follicle stimulating hormone/luteinizing hormone releasing factor—FSH/LH-RF* (see Unit 16, Figure 46)

follicle stimulating hormone/luteinizing hormone releasing factor (FSH/LH-RF)*

Neurons that produce FSH/LH-RF are scattered throughout the hypothalamus. Figure 7 shows the position of the major *nuclei* (groups of neurons) in the hypothalamus. If neurons are destroyed in those areas that are shaded grey in the Figure, then production of FSH and LH is diminished.

hypothalamic nuclei*

FSH/LH-RF is not species specific in structure and will both cause the release, and stimulate the synthesis, of FSH and LH in the anterior pituitary. The activity of FSH/LH-RF is very quickly over, because it is rapidly degraded in the blood and excreted by the kidneys. The hypothalamus may also produce an FSH/LH-

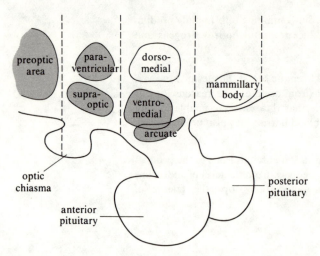

RIF—a release *inhibiting* factor—which suppresses the release of FSH and LH, possibly by blocking FSH/LH-RF receptors in the anterior pituitary. Such 'blockers' (antagonists) of FSH/LH-RF have also been synthesized, and their usefulness in birth control is something we return to in Section 5.

So the control of gonadal function is a three-tier system between the hypothalamus, the pituitary and the gonads. The extra tier, the hypothalamus, brings with it a site for the regulation of reproduction via other areas of the brain. The importance of this relationship will become clear in Section 3.

Before we look at the precise control of the oestrous cycle by hormones, there is one other pituitary hormone to mention—*prolactin*. This is a peptide (relative molecular mass 25 000) produced by cells in the anterior pituitary, and its release is also under hypothalamic control. There is evidence for both a *prolactin releasing factor* (*PRF*) and a *prolactin release-inhibiting factor* (*PIF*). We look at the role of prolactin in more detail in Section 2.5.

prolactin*

prolactin releasing factor (PRF)*
prolactin release inhibiting factor (PIF)*

2.2 Fine control of the oestrous cycle

In the last 10 years, the development of the technique of radioimmunoassay (Unit 16, Section 6.1) has made it possible to measure accurately the levels of hormones circulating in the blood during the oestrous cycle. Figure 8 shows the relative changes in FSH, LH, oestradiol and progesterone in the normal human

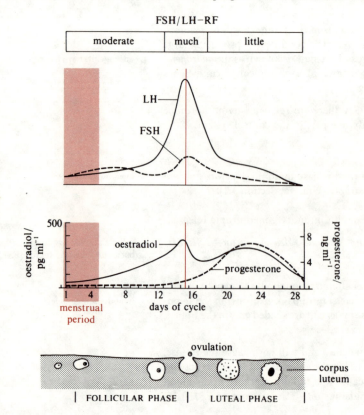

FIGURE 8 The changes in gonadotropins (FSH, LH) and steroids (oestradiol and progesterone) during the menstrual cycle. The changes in the ovary are also indicated.

menstrual cycle. Note that both FSH and LH peak at ovulation, that the oestradiol level rises just before the FSH and LH peaks, and that progesterone rises after ovulation. The question is: are these changes related to one another, and if so, how?

During the *follicular phase of the menstrual cycle*, the low level of circulating oestrogens (from the ovary), results in an increase in the secretion of FSH from the pituitary.

follicular phase of cycle*

☐ What sort of feedback does this illustrate?

■ Negative feedback.

FSH stimulates follicular development in the ovaries if LH is present. Growth of the follicles results in an increased secretion of oestrogens from the follicular cells. Oestrogens directly influence follicular growth and also make the ovaries more *responsive* to the *gonadotropins* LH and FSH. As the level of oestradiol increases during the late follicular phase just before ovulation, there is a marked *positive* feedback effect of oestradiol on gonadotropin release resulting in the peaks (*surges*) in LH and FSH seen in Figure 8. This combined rise in gonadotropins brings about the final maturation of the follicle resulting in ovulation.

surge (phasic release) of gonadotropin*
ovulatory phase of cycle*

The high level of oestradiol now exerts a *negative feedback* effect on gonadotropin release (FSH and LH fall), and consequently the level of oestradiol (from the ovary) falls. The corpus luteum in the ovary (under the influence of LH) is responsible for the massive secretion of progesterone, which peaks 5–6 days after ovulation, and also for the second rise in oestradiol, which begins just after ovulation. If fertilization does not occur, the corpus luteum regresses, and consequently both progesterone and oestradiol levels fall. A new cycle then begins with menstruation.

luteal phase of cycle*

This sequence of events is typical of the human menstrual cycle. In other mammals, slight variations can occur. For example, whereas progesterone appears to play no part in ovulation in the human, it is known to induce ovulation in cows, rabbits, rats and sheep. The maturing follicle in the ovary appears to be able to secrete small quantities of progesterone just before ovulation.

This pattern of hormonal changes during the menstrual cycle raises a number of questions. First, how can FSH and LH be involved with follicular growth when their levels are so consistently low during this early period of the cycle? It is possible that the *cumulative* effect of these gonadotropins on the ovary is the important factor in follicular development.

☐ There is an alternative possible explanation. Can you think of it? (So far, we have a messenger molecule explanation, but what about changes in the ovary itself?)

■ It is possible that receptors for LH and FSH on the follicles undergo a degree of sensitization with time and therefore respond more strongly to the same level of gonadotropins later in the cycle. (Receptor sensitization is discussed in Unit 16, Section 8.)

Another puzzling aspect of hormonal changes during the menstrual cycle is that oestradiol appears to have both stimulatory and inhibitory effects on gonadotropin levels at different times. Oestradiol normally has a negative feedback effect on gonadotropin release, and this results in the sharp drop in FSH and LH just after ovulation. Progesterone augments the inhibitory effects of oestradiol on gonadotropin secretion. Low levels of oestradiol during the follicular phase allow FSH and LH secretion, which, in turn, promotes follicular maturation and the secretion of oestradiol from the follicle. The site of this negative feedback is thought to be on the production and release of FSH/LH-RF in the hypothalamus.

The target for the positive feedback effect of oestradiol just before ovulation may be either at the hypothalamus or on the pituitary itself. In primates, the pituitary seems to be involved. Here, cells that secrete FSH and LH become *more sensitive* to FSH/LH-RF from the hypothalamus. The changing sensitivity of the anterior pituitary is a direct effect of rising levels of oestradiol during the follicular phase of the cycle.

In other mammals, the main site (or sites) of the feedback actions of steroids may be different. If the anterior three nuclei of the hypothalamus in the *rat* are destroyed (or artificially stimulated), ovulation is prevented (or initiated). Similarly, if oestrogen is implanted in this area of the hypothalamus, ovulation results. While this area of the hypothalamus appears to be concerned with ovulation, the other area of the hypothalamus that promotes gonadotropin release (the nuclei just above the pituitary, see Figure 7) is involved simply with gonadal maintenance. Therefore these contradictory effects of feedback regulation of oestrogen on gonadotropins can be explained, at least in the rat, by the existence of *two* sites in the hypothalamus; one that responds by positive feedback to oestrogens (the anterior areas—known as the *surge* centre) while the other responds by negative feedback to oestrogens (the areas above the pituitary—known as the *tonic* centre). This idea is summarized in Figure 9.

In women, unlike other mammals, it is usual for one follicle at a time to develop to the stage when it responds to the surge of LH by ovulating, although earlier in the cycle a number of follicles may have begun to develop under the influence of FSH. The sensitivity of the one follicle is probably due to its ability to secrete more oestrogen than the others. The *'fertility drugs'* that became notorious and newsworthy in the late 1960s because they led to quintuplet and sextuplet births were gonadotropins. Some fertility drugs cause many follicles to reach maturation at the time of ovulation and their simultaneous release provides multiple targets for fertilization.

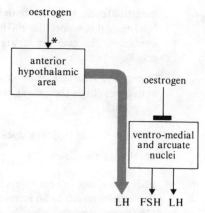

FIGURE 9 The different effects of oestrogen on the two main sites of FSH/LH-RF production in the hypothalamus.

2.3 The corpus luteum—maintenance and luteolysis

From Section 2.2, it should be clear that the complete cycle has three stages: a follicular stage, an ovulatory stage, and a luteal stage. Some animals, for example primates, have all three stages. After ovulation, the ruptured follicle changes shape and becomes a corpus luteum. Under the influence of LH, the corpus luteum starts to produce oestradiol and progesterone. If fertilization and implantation are successful, the corpus luteum persists in the ovary.

Other mammals do not show the three stages of the cycle clearly. In rats, if mating does not occur, the corpora lutea produce very little progesterone and the uterus does not undergo the changes associated with high levels of progesterone (e.g. vascularization). If mating occurs, stimulation of the vagina and cervix by the penis is communicated via nerves to the hypothalamus, and this promotes the release of prolactin from the anterior pituitary.

☐ What mechanism is employed between the hypothalamus and the pituitary?

■ To promote prolactin release *either* the hypothalamus must secrete a PRF *or* there must be a reduction in the release of an inhibiting factor (PIF) from the hypothalamus.

In the rat it seems that PIF is blocked. The release of prolactin promotes cholesterol metabolism in the corpora lutea. This is not quite the full story. LH is also important in the conversion of cholesterol to progesterone, at least for the first two weeks of gestation. After that, the placenta maintains the pregnancy by production of a *luteotropic factor* that sustains the manufacture of progesterone by the corpora lutea.

luteotropic factor

In yet a third group of mammals (e.g. the cat and ferret), there is clearly only one stage of the cycle, and that is the process of follicular development. These mammals are called induced ovulators because they ovulate in response to mating. We shall return to them in Section 3.

If pregnancy does not result, the corpus luteum undergoes *luteolysis* and the production of progesterone ceases (see Figure 8), but if the uterus is removed (hysterectomy), then the corpus luteum is maintained in some species.

☐ What can you deduce from this effect of hysterectomy?

■ The uterus is probably producing a substance that causes the corpus luteum to break down (a *luteolytic factor*).

luteolytic factor

This substance appears to be a *prostaglandin* ($PGF_{2\alpha}$—see Unit 16, Section 6.1). Injection of $PGF_{2\alpha}$ into pregnant females causes luteolysis and subsequently abortion. $PGF_{2\alpha}$ has been used to induce abortion in women, and it appears to do so by causing contraction of the uterine muscles.

prostaglandin $PGF_{2\alpha}$

14

□ If prostaglandins can cause abortion, what must happen to $PGF_{2\alpha}$ levels in pregnancy?

■ A signal, presumably from the fetus, must suppress the synthesis and release of $PGF_{2\alpha}$ from the uterus (an anti-luteolytic signal).

So, to maintain pregnancy, the corpus luteum must produce progesterone (at least for the first few months), and this ensures implantation of the blastocyst in the uterus and the inhibition of luteolysis. At term, or if fertilization does not occur, luteolysis must happen to initiate a new cycle.

2.4 Hormones and male reproduction

In the same way that the ovary is under the control of both the hypothalamus and the pituitary, so too is the testis. And just as females can be sexually active either continuously or seasonally so too can males. In most species, the breeding behaviour of the male coincides with that of the female.

The important components of the testis are the seminiferous tubules (which produce sperm) and the Leydig cells (which secrete androgen). Both FSH and LH affect the testis. Growth of the testes is clearly under gonadotropic control. A rapid increase in testicular mass occurs at puberty, and this is correlated with an increase in plasma (blood) levels of FSH and LH. Administration of LH to adult mammals that have been hypophysectomized results in growth of the Leydig cells and an increase in androgen secretion. LH probably affects the synthesis of androgen at some step between cholesterol and pregnenolone. LH affects the activity of a number of enzymes in the steroid synthesis pathway, and therefore probably has a general tropic action on Leydig cell metabolism and protein synthesis. LH produces a rise in cyclic AMP levels in gonadal cells, which results in increased synthesis of steroids (see Figure 10).

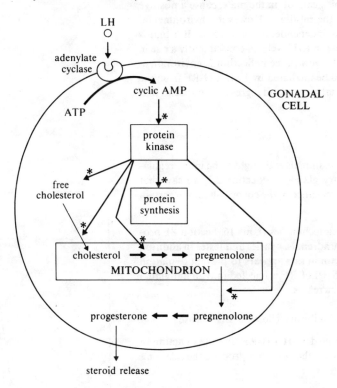

FIGURE 10   The effect of LH on gonadal cell metabolism. Cyclic AMP activation of protein kinase promotes steroid synthesis and release. As you can see, protein kinase has several effects on the route to synthesis of steroids.

Hormonal control of spermatogenesis is a little more complex. FSH can produce a very significant increase in the diameter of the seminiferous tubules, but both LH and FSH are essential for completion of spermatogenesis. The major site of FSH action appears to be on Sertoli cells, where membrane receptors for FSH have been identified. Sertoli cells also have receptor sites for androgens.

□ Where would you expect to find these androgen receptors (i.e. in or on the cell?)

■ Androgens are steroids, so the principal binding sites would be on cytoplasmic proteins and not on the cell membrane.

Sertoli cells therefore have receptors for both peptide *and* steroid hormones.

The feedback control of FSH and LH production exerted by the testes appears not to involve androgens (see Figure 11). Very high levels of androgens are necessary before there is any feedback inhibition of LH and FSH. However, small amounts of oestrogen immediately depress gonadotropin secretion in males, and as we saw earlier (in Section 2.1), oestrogen is a normal product of the testes.

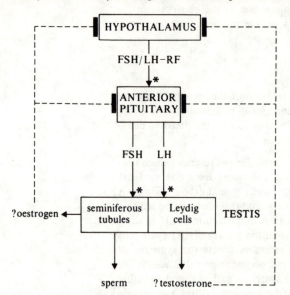

FIGURE 11 Feedback control of releasing factor and gonadotropin release is brought about possibly by testosterone or by oestrogen.

FSH/LH-RF is found in the male hypothalamus, but there is little evidence for a cycle, such as the oestrous cycle, in males. So what controls the release of FSH/LH-RF? Unlike in the female, steroids cannot, in the male, cause a positive feedback effect at the hypothalamus or the pituitary. However, environmental changes can regulate gonadotropin levels independently of steroids. If a bull is placed close to a cow in heat, there is a rise in LH level and consequently a rise in testosterone production from the testes. As we will see in Section 3, environmental effects on steroid production appear to be mediated by FSH/LH-RF from the hypothalamus, and this can explain gonadal growth and regression in species that are seasonally active.

2.5 The role of prolactin

Classically, the major role of prolactin in mammals is thought to be in the regulation of milk secretion from the mammary glands. In Section 2.3, we saw that prolactin may also be involved in the maintenance of the corpus luteum in certain mammals.

With the development of receptor-binding techniques (Unit 16, Section 8), prolactin binding sites have been found in the adrenals, ovaries and testes in addition to mammary tissue. High levels of prolactin in rats appears to prevent ovulation by suppressing the surges of LH. When FSH/LH-RF is injected into an individual with high prolactin levels, LH secretion increases.

□ Where must prolactin be causing this inhibitory effect on LH release?

■ If LH levels change in response to injected FSH/LH-RF, the site of action of prolactin must be in the hypothalamus, on the cells that produce the releasing factor.

During a normal oestrous cycle, the levels of prolactin peak at the time of ovulation (Figure 12), probably in response to the rise in oestrogen level; if the ovary is removed, prolactin levels decrease and the mammary glands regress. The rise in prolactin may serve two functions: first, it is present ready for the luteotropic function just after ovulation, and second, it may terminate the ovulatory surge in gonadotropins, acting therefore as a fine regulator of ovulation.

There is no clear *menstrual* rhythm in prolactin level unlike the oestrus peak in the rat. The 'anti-ovulatory' effect of prolactin in women would be important during the suckling period and may well be responsible for the irregular cycles that occur after a birth. In the !Kung tribe in Africa, no methods of birth control are

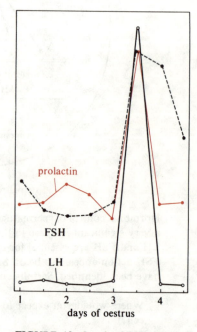

FIGURE 12 Levels of FSH, LH and prolactin during a 4-day oestrous cycle in the rat. Ovulation is during day 3.

16

practised, and yet women have a child relatively infrequently—once every 4 years. The key to the relatively small number of pregnancies seems to be that children are suckled from birth to 3 years of age. Moreover, the children are allowed free access to the nipple, and this high suckling frequency maintains high levels of prolactin. In western societies, breast feeding is usually of short duration and at regular times, and therefore not a reliable method of contraception. Some women have a persistently high level of prolactin in the blood. Associated with this is the absence of LH surges and so lack of ovulation. It appears that in women, the absence of LH surges may be due to a decreased sensitivity of the pituitary to FSH/LH-RF rather than a lack of this releasing factor.

This condition is now easy to treat, mainly because the control of prolactin secretion is better understood. The involvement of neurotransmitters in the control of pituitary function has been suspected for a very long time. In Section 2.1, we saw that the release of prolactin appears to be under the control of both a PRF and a PIF from the hypothalamus, probably with the PIF control dominant (i.e. prolactin release is normally inhibited). If levels of the neurotransmitter, *dopamine*, are raised in the hypothalamus or pituitary, prolactin secretion is inhibited; if the release of dopamine is prevented in some way, then prolactin secretion is stimulated. Neurons that release dopamine have been identified in the hypothalamus (Figure 13), and dopamine appears to be the proposed PIF. Prolactin itself can stimulate the release of dopamine.

dopamine*

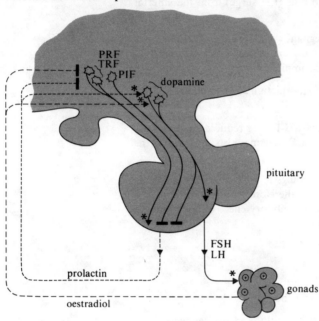

FIGURE 13 The effects of PRF, TRF, PIF and dopamine, released from neurosecretory cells in the hypothalamus on FSH, LH and prolactin release from the anterior pituitary. Note the feedback effects of prolactin and oestradiol on the hypothalamus.

☐ What would the consequence be of this feedback?

■ Because dopamine blocks prolactin release, this feedback system would be a negative one and prolactin levels would drop.

The treatment of women who are not ovulating because of a high prolactin level in the blood involves not dopamine itself but an agonist (see Unit 16, Section 8). Daily treatment with a dopamine agonist lowers prolactin level, and ovulation and menstrual cycles return within two months.

This involvement of neurotransmitters with the secretion of hypothalamic and pituitary hormones opens up a new and exciting field. The effect of dopamine on prolactin secretion is not the only example. Noradrenalin can stimulate the release of FSH and LH; *serotonin* inhibits LH and appears to control the rhythmic secretion of many pituitary tropins (e.g. ACTH, TSH, LH and FSH), and dopamine can stimulate the release of FSH/LH-RF. The sites of interaction of these transmitters with the neurosecretory terminals may be both in the hypothalamus and in the pituitary. This interaction of the nervous system with neurosecretory neurons presents possibilities of feedback control via a short loop and could provide the mechanism for slow (continuous) release and fast (cyclical) release of the pituitary gonodotropins.

serotonin

This close relationship between the nervous system and the endocrine system, a continuing theme in Units 16, 17 and 18, is especially important in the timing and synchronization of reproductive cycles, as we see in the next Section.

Summary of Section 2

1 Two pituitary gonadotropins, FSH and LH, (present in both sexes) promote gonadal steroid production and the maturation of gonadal tissue.

2 The ovary and the testis produce steroid hormones (oestrogens, androgens and progesterone), which, via a feedback loop, control gonadotropin release.

3 The hypothalamus controls gonadotropin release by an FSH/LH releasing factor (FSH/LH-RF).

4 The menstrual cycle proceeds through a number of phases: menstruation, follicular growth phase, ovulation, luteal phase.

5 FSH promotes follicular growth; LH initiates ovulation and the formation of a corpus luteum which secretes progesterone.

6 During the follicular phase, low levels of circulating oestrogens promote (via negative feedback) secretion of LH and FSH, and rising oestrogen levels promote an LH surge and ovulation. During the luteal phase, oestrogens and progesterone suppress (via negative feedback) the secretion of FSH and LH.

7 Two separate areas of the hypothalamus have been proposed to release FSH/LH-RF—one involved in the tonic (constant) release of gonadotropins and the other in the phasic release of gonadotropins (LH surge) from the anterior pituitary.

8 The corpus luteum persists during pregnancy and produces progesterone which is essential for the maintenance of a receptive uterus. Prostaglandin ($F_{2\alpha}$) breaks down the corpus luteum if there is no implantation.

9 FSH and LH promote testicular androgen production and spermatogenesis in the male.

10 Prolactin prevents the production of FSH/LH-RF (or reduces the sensitivity of the pituitary to it), thus preventing ovulation. Prolactin is also important in the preservation of the corpus luteum.

11 Neurotransmitters released from nerve endings in the hypothalamus locally control the production of releasing factors, and therefore pituitary hormones, or act directly on the anterior pituitary (e.g. the inhibitory effect of dopamine on prolactin release from the anterior pituitary).

Objectives and SAQs for Section 2

Now that you have completed this Section, you should be able to:

★ describe the major endocrine changes that control the menstrual cycle.

★ demonstrate an understanding of various aspects of the hormonal control of the mammalian reproductive cycle by predicting the results of various experimental procedures involving hypophysectomy and/or hormone treatment.

★ give evidence for a hypothalamic–pituitary–gonadal link in the control of the menstrual cycle.

★ distinguish between negative and positive feedback control of gonadotropin release.

★ describe the functions of the corpus luteum, and distinguish between luteotropic and luteolytic factors.

To test your understanding of this Section, try the following SAQs.

SAQ 3 (*Objectives 4 and 6*) What effect would you predict from the continuous administration of progesterone to a female monkey (a menstruator) for at least 90 days at a dosage:

(a) such that the level in the blood was at least that found normally on day 23 of the cycle (see Figure 8)?

(b) such that the level in the blood was just 10 per cent of that on day 23?

Consider the effect on the ovaries, the endometrium and on blood levels of the naturally secreted steroids.

SAQ 4 (*Objective 4*)   If a male rabbit were hypophysectomized, which of the following hormones—FSH, LH, progesterone, testosterone—would you need to inject to maintain the entire reproductive system in a fully functional condition for (a) 2–3 weeks and (b) indefinitely?

SAQ 5 (*Objective 5*)   If the gonadotropin LH was injected into a mature male mammal, which one of the following results would you expect?

(i) an increase in oestrogen;

(ii) an increase in testosterone;

(iii) a decrease in the activity of the testes;

(iv) a 28-day cycling of male hormones;

(v) increased sperm production by the testes.

SAQ 6 (*Objective 7*)   From statements (i)–(vi), choose that which accurately describes the events following fertilization of the zygote in (a) a rat and (b) a woman.

(i) The fertilized egg implants in the uterus and begins to secrete progesterone.

(ii) After implantation, the ruptured follicle forms a corpus luteum, which starts to produce progesterone under the influence of prolactin.

(iii) After implantation, the corpus luteum breaks down.

(iv) After implantation, the corpus luteum starts to produce progesterone under the influence of a prostaglandin.

(v) After implantation, the corpus luteum starts to produce progesterone under the influence of the gonadotropin LH.

(vi) After implantation, the corpus luteum starts to produce oestrogen, which maintains the endometrium.

SAQ 7 (*Objective 5*)   Construct a simple flow diagram to show how prolactin secretion, initiated by suckling, can act as a contraceptive. (By flow diagram, we mean a diagram like Figure 11).

SAQ 8 (*Objectives 5 and 6*)   Construct a flow diagram to show the hypothalamic–pituitary–ovary links in the control of the menstrual cycle.

3   Environmental influences on the timing of reproduction

The TV programme Only in the Mating Season is particularly relevant to Section 3.1.

In Section 2, we looked at the endocrine changes in the different phases of the menstrual cycle. In *continuous breeders* (Table 1), the end of one cycle is marked by luteolysis and then the breakdown of the uterine lining. New follicles ripen under the renewed influence of FSH and the cycle begins again. Most mammals however are not continuous breeders but show a limited number of cycles during a certain period of the year (i.e. they are *seasonal breeders*). So, in addition to describing the control mechanisms involved in one cycle, we have to seek other mechanisms that *initiate* the breeding season.

Synchronization of the reproductive state of both sexes maximizes the chances of fertilization. As the subsequent processes of implantation and the length of gestation are 'fixed' in terms of duration for any given species, to improve the chances of survival of the newborn, the precise timing of fertilization in the year is crucial. Between them, mammalian species illustrate a variety of strategies resulting in the optimal timing for birth; these strategies include delayed implantation and seasonal receptiveness. Within a breeding season, species may be polyoestrous or monoestrous. The process of 'coming into season' involves *environmental cues* such as the change in day length, and the timing of reproductive encounters involves visual, olfactory and auditory (sight, smell and sound) cues from other individuals.

timing of reproduction, control by nerves and external factors

environmental cues*

3.1 Light and temperature cues

The distinction between continuous and seasonal breeders is not a rigid one. Most mammals show peaks of reproductive activity. Even our own species shows peaks of conception in May and June in the Northern Hemisphere, and in November and December in the Southern Hemisphere. Generally, seasonal breeders are those species in which the gonads regress completely, and so become inactive, out of the breeding season.

Depending on the latitude, both *day length* (number of hours of light per 24 hour day) and temperature vary throughout the year. In the Northern Hemisphere, day length increases from December to June and then gradually decreases. Average monthly temperature increases steadily from February to August. Depending on the size and length of gestation of a species, sexual behaviour occurs in the spring or in the autumn.

day length*

☐ What would the advantage of autumn mating be in a species (e.g. sheep) that has a long gestational period?

■ A long gestational period, extending over the winter, means that the young are born in the following spring coincident with increasing food supply. (If you look again at Table 1, you can see that species size and gestation period are fairly closely correlated).

Figure 14 shows the periods of oestrus of two groups of Suffolk ewes subjected to different lighting conditions. In the control group, as the day length shortens, oestrous cycles are initiated in September. In the experimental group, with a seasonally reversed day-length cycle begun in the first year, the initiation of the second set of oestrous cycles is in May.

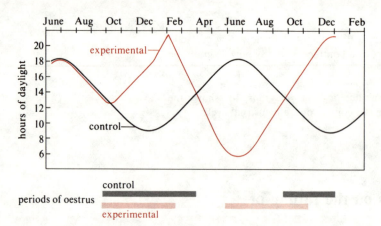

FIGURE 14 The sexual season of two groups of Suffolk ewes subjected to different lighting conditions. The sexual season (oestrus) in each case is shown as a solid band beneath the corresponding light curve. In all cases, oestrus started 10–14 weeks after the change-over from increasing to decreasing hours of daylight.

☐ To what specific cue do the ewes appear to respond?

■ Decreased day length; as the hours of daylight decrease, oestrous periods are initiated.

This shows that where all the other effects are left as they normally are, simply imposing a reversed light cycle can entirely rephase the breeding season. This is also true of deer. These two species are known as *short-day breeders*. Mammals such as the horse (with a 10 month gestation period) are *long-day breeders*, coming into 'season' in the spring. In this Unit, we use short-day and long-day breeders to indicate those species in which mating is initiated by an increase in, or decrease in, day length. Some species (e.g. foxes), with a wide north to south distribution, may come into breeding condition at different times of year (earlier in the south), but they are still responding to an increase in day length.

short-day breeders*
long-day breeders*

Figure 15 shows some effects of *photoperiod* (day length) on Soay rams. Note that the change from long days (LD) to short days (SD) stimulates FSH secretion and a growth in testis size. LH levels show episodic secretion and testosterone level gradually rises. When the rams are transferred back to a long-day schedule 18 weeks later, FSH and LH secretion falls and the testes regress. How can changes in the photoperiod trigger gonadotropin secretion from the pituitary? There are a number of possible routes, and all inevitably lead to the hypothalamus and the level of FSH/LH-RF.

photoperiod, photoperiodicity*

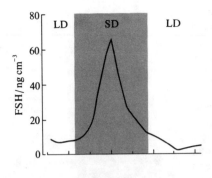
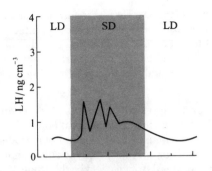
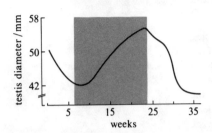
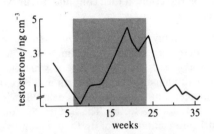

FIGURE 15 Change in testicular size and levels of various hormones in the blood of adult Soay rams transferred from periods of exposure to long days (LD—16 hours light and 8 hours dark) to short days (SD—8 hours light and 16 hours dark) at week 6. Animals were maintained on short days until week 24 when they were returned to long days.

In rodents, the eyes and the *pineal gland* of the brain have been implicated in breeding cycles. If a hamster is blinded on long days, there is immediate regression of the gonads, just as if it had been transferred to short days. If the pineal gland is removed, this regression does not occur. If the pineal is removed from hamsters kept under a short-day regime, the gonads grow.

pineal gland*

☐ What does the pineal probably secrete on short days?

■ An anti-gonadotropic substance.

On long days, the pineal does not exert an inhibitory effect. The pineal gland secretes an amine called *melatonin*. Sometimes melatonin causes gonadal regression, and at other times it stimulates gonadotropin release.

melatonin*

Figure 16 shows the effects of melatonin on the mass ('weight') of the testes in four species of rodents. The testes of the grasshopper mouse and hamster regress when exposed to melatonin, but those of the house mouse and albino rat appear to be unaffected. The two melatonin-sensitive species are seasonal breeders, whereas the two melatonin-insensitive species breed continuously. Melatonin from the pineal gland seems to be a significant inhibitor of gonadal function in seasonal breeders. One possible explanation of the mode of action of melatonin is that it regulates the production of an as yet unidentified anti-gonadotropin in the pineal gland. Conversely, melatonin may act directly on the hypothalamus, or on the pituitary, or directly on the gonads.

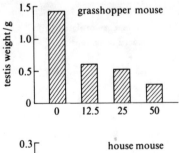
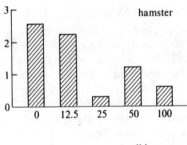
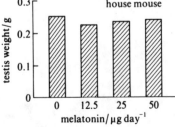
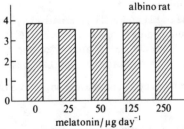

FIGURE 16 The effects of melatonin (after 60 days administration) on testicular size in four species of rodent. Grasshopper mouse and hamster are seasonal breeders; house mouse and albino rat are continuous breeders.

The pineal produces melatonin in a cyclical fashion, in other words there is a 24 hour fluctuation in melatonin level. The effect of giving melatonin to an animal depends very much on *when* it is given in the day. Here, therefore, is a substance in the blood, produced by the pineal, whose effects vary depending on the time of day at which the brain is exposed to it. Because the secretion of melatonin may depend on the onset of darkness, this is clearly an important physiological mechanism by which the pineal 'tells' the brain whether it is in a 'long day' or a 'short day'.

The role of the photoperiod is probably as a 'coarse tuner', synchronizing development of the reproductive system in members of a local population, both male and female. Some mammals have 'in-built' yearly rhythms of reproduction, for example some hibernating species. Light may act simply to adjust an in-built 'clock' in the hypothalamus. Once this 'clock' (which could be the production and secretion of FSH/LH-RF) is set to the right time, it then continues to run without the necessity for another light cue.

Photoperiodicity may do more than simply prime ovulation. In some species, there is a delay between fertilization of the egg and subsequent implantation of the blastocyst—*delayed implantation*. This means that some species which mate in spring may delay implantation until autumn and then give birth in the following spring. Hedgehogs mate in late summer, but implantation is delayed, with the result that the young are born in the following spring. The badger, the roe deer, the mink and the marten can all delay implantation (for up to four months in the deer): this delay can be terminated if the photoperiod or the temperature are changed.

delayed implantation*

Rats and mice also show delayed implantation. Females ovulate 24 hours after giving birth. If copulation occurs at this time, the subsequent gestation period is up to 40 days long instead of the normal 21. If, however, lactation is prevented, the gestation period reverts to 21, or if oestrogen is given during lactation, the blastocysts implant. This is interpreted as an *oestrogen* requirement for implantation in the rat; during lactation, this oestrogen is lost by being secreted in the milk.

The implantation mechanisms of wallabies and kangaroos are even more puzzling. In the wallaby, implantation will occur even if the pituitary is removed. Wallabies and kangaroos, like the rat, undergo ovulation after giving birth. The blastocyst, it seems, does not implant immediately but survives for many months. If the 'joey' in the pouch dies during this time, or if it stops suckling, the blastocyst will then implant and start to develop.

In some animals, for example birds, sea turtles and bees, the female can store sperm in the reproductive tract for long periods of time (a few years in the turtle and seven years in a queen bee). One case of sperm storage has been reported in mammals—in bats. Some bats copulate in autumn but do not *ovulate* until spring, and this was interpreted to mean that sperm were stored over the winter. Closer behavioural studies have now shown that bats copulate again in the spring, and so fertilization is probably achieved with fresh sperm.

3.2 Cues from other individuals

While many mammals are *spontaneous ovulators*, that is they ovulate in response to a surge in gonadotropin release from the pituitary, others are *induced ovulators*. In the cat, rabbit, ferret, mink and some species of mice, courtship and copulation actually stimulate ovulation. Stimulation of the cervix and the vagina promotes, through the nervous system, the release of FSH/LH-RF from the hypothalamus. The evidence for this role of the nervous system comes from removing the pituitary gland at varying times after copulation. If the pituitary is removed up to 1 hour after copulation, then no ovulation will occur. Beyond 1 hour after copulation ovulation occurs even in the absence of the pituitary; by this time, the hypothalamus has initiated the release of LH and FSH. Alternatively, if the nerve pathway from the vagina to the hypothalamus is blocked within one minute of coitus having taken place, ovulation does not occur; a delay of one minute is sufficient to allow the copulation 'signal' through to the hypothalamus.

spontaneous ovulators*
induced ovulators*

Induced ovulators do not have an oestrous cycle comparible with those of spontaneous ovulators (Figure 17).

☐ Why might induced ovulation offer a selective advantage over spontaneous ovulation?

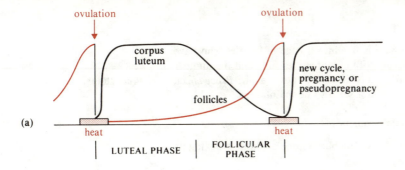

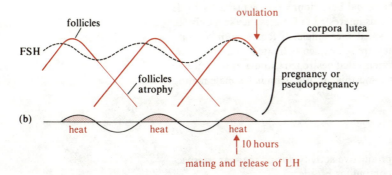

FIGURE 17 The chain of ovarian events in spontaneous and induced ovulators is basically the same. (a) Spontaneous ovulation is cyclical, following specific periods of heat in the oestrous cycle. (b) Induced ovulation occurs as a result of stimulation of the cervix during copulation. Induced ovulators experience alternating oestrous and anoestrous periods, and theoretically are sexually receptive at all times.

■ At every mating, sperm and ova are likely to meet in induced ovulators, whereas in spontaneous ovulators, many matings will be too early or too late for fertilization of the ova.

Induced ovulation is a common feature of mammals that live singly in field or forest. The more or less continuous state of sexual readiness provides a good chance that, at any occasional meeting, the female will be receptive. Not all solitary species are induced ovulators; in some species, seasonal 'gatherings' of individuals improve the chances of reproductive success.

Mammals show induction of other phases of the cycle in addition to the ovulatory phase. In Section 2.3, we saw how the luteal phase could be induced.

☐ In which species was this?

■ In the rat; stimulation of the vagina during copulation promotes the maintenance of the corpora lutea by stimulating the release of prolactin.

The hamster and several species of mice also show such an induced luteal phase.

Physical contact is not always necessary to produce these preparatory hormonal responses. That complex displays (involving colourful plumage) or the mere sight of a finished nest may synchronize reproductive states in birds is well known, but are there parallels in mammals? Can an individual communicate its reproductive condition to another individual and thereby affect that second individual? There are a number of possible means of communication: by odour, visual displays or sound. In birds, the latter two seem to be employed, and in mammals, odours are also important.

Female rats, in constant oestrus (produced by constant light exposure), induce oestrus in normally cycling females via an olfactory cue. In mice, if females are housed together, and then each is allowed access to a male, there are unexpected results.

☐ Given that the mouse oestrus cycle is 5 days long, with one day on heat, how many mice would you expect to mate on each night?

■ You would expect them to be on heat and therefore mate randomly, so this means a fifth of the total on each night.

But now look at Figure 18. Almost 50 per cent of the females mated on the third night! This effect of females synchronizing oestrus was described first by *Whitten* who ascribed it to an odour (a *pheromone*).

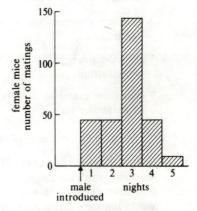

FIGURE 18 The Whitten effect: the synchronization of oestrus in groups of mice.

Whitten effect (synchronization of oestrus)*
pheromones*

If a male of a different strain of mouse is put into a cage of females who have recently been mated, the blastocysts do not implant successfully and the number of pregnancies is fewer than normal. In this situation, known as the *Bruce effect*, the corpora lutea of the first mating do not seem to be maintained.

Bruce effect*

☐ What could the action of the male odour be?

■ Prolactin is luteotropic in the mouse, so the pheromone could be blocking its release.

Pheromones are also important in the sexual behaviour of some primates. Male monkeys fitted with nasal plugs are not as interested in females they can only see as are male monkeys that can also smell the female. If female monkeys are castrated, no mating occurs, but if they are subsequently treated with oestrogen, the mating frequencies rise sharply. It seems that oestrogen stimulates secretion of certain kinds of acids from the vagina, and this, in turn, stimulates male mating behaviour.

Summary of Section 3

1 Most mammals show peaks of reproductive activity during the year irrespective of whether they are seasonal or continuous breeders.

2 Light is implicated in the initiation of a breeding season; some mammals respond to decreasing day lengths while others respond to increasing day lengths.

3 The perception of day length in seasonal breeders appears to involve the pineal gland in the brain, which secretes melatonin. This amine causes regression of the gonads, possibly through an anti-gonadotropin.

4 Some mammals have the ability to delay the implantation of a fertilized egg.

5 Cues from other individuals are important in the regulation of the reproductive cycle. These cues may be tactile, visual or olfactory. Pheromones synchronize oestrous cycles in female rats and mice, and female pheromones induce sexual activity in male monkeys.

6 Some species are induced ovulators and respond to the tactile stimulation of copulation. This stimulation is conveyed to the hypothalamus via the nervous system. A surge in LH results, and ovulation occurs.

Objectives and SAQs for Section 3

Now that you have completed this Section, you should be able to:

★ give two examples where the nervous system plays a role in the timing of reproduction.

★ describe three pieces of evidence for external influences on the timing of reproduction.

★ give three examples that illustrate the synchronizing, by individuals, of their reproductive state.

To test your understanding of this Section, try the following SAQs.

SAQ 9 (*Objectives 8 and 9*)   If a mature doe rabbit was hypophysectomized and soon afterwards put into a cage with a male, LH would have to be given to make fertilization possible. If you repeated the experiment, this time using mature rats, would the same action allow fertilization to take place?

SAQ 10 (*Objectives 8 and 9*)   What sort of investigations would you perform to determine whether a mammal was an induced rather than a spontaneous ovulator? Is an induced ovulator affected by environmental factors so far as the cycle of events in its ovary is concerned?

SAQ 11 (*Objectives 8 and 9*)   Farmers have known for many years that, in some cases, ewes may be brought into oestrus earlier than otherwise by penning them next to a ram. This also has the effect of tending to synchronize their cycles. Suggest a plausible explanation for this phenomenon.

4 Sex determination—the action of steroids

Before you begin this Section, it would be useful to look at Unit 11, Section 4.1, to remind yourself of the key episodes in development.

In mammals, haploid gametes, produced by meiotic division from diploid cells, combine to form a diploid zygote. The genotypic sex of a mammalian fetus is determined at fertilization by the presence of an XY (male) or an XX pair of chromosomes (Figure 19).

sex determination*

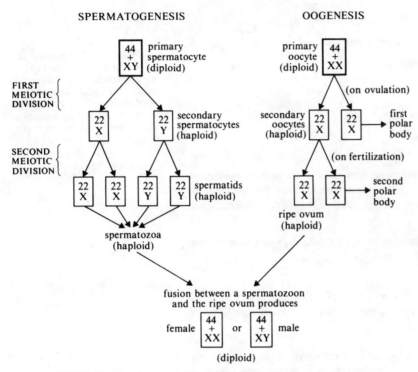

FIGURE 19   The chromosomal events in the production of spermatozoa (spermatogenesis) and of a ripe ovum (oogenesis).

But is the Y chromosome the only thing that determines 'maleness'? Certainly it appears so when clinical disorders are considered. If, due to chromosomal problems, fertilization results in an individual with 47 (XXY) or 48 (XXXY) chromosomes, then a testis will develop, but if the individual is XO, then it is phenotypically female. In recent years, however, it has become clear that genetic sex is much more complicated than simply the presence or absence of the Y chromosome. Genes essential for male development are located on the X as well as the Y chromosome, and genes essential for the development of both male and female characteristics are located on other chromosomes.

In the human embryo, the gonads are not recognizably of either sex until 6 weeks after fertilization. Then differentiation begins, and the secretion of steroids from the gonads determines their subsequent development. If the gonad secretes testosterone, a male reproductive system will develop. If no testosterone is present, then the female ducts and genitalia develop. The ability of the fetal testis to synthesize testosterone coincides with the differentiation of the Leydig cells. Testosterone plays two vital roles: first, it is essential for the maturation of the seminiferous tubules and for spermatogenesis. Second, testosterone released into the fetal circulation is involved in promoting 'maleness' in other tissues. In the adult, androgen synthesis is under the control of gonadotropins (Section 2.4), and this appears to be true in the fetus. The site of gonadotropin release could be the fetal pituitary or the placenta, and gonadotropin receptors have been demonstrated on the fetal testis. Similarly, the fetal ovary can secrete oestrogens at this early stage, but whether they are important in the development of the female fetus is, as yet, unclear.

phases of mammalian development, effects of steroids on*

If testosterone serves as the fetal androgen, are its molecular mechanisms the same as in the adult? In the adult, testosterone is thought to have few effects of its own, but it serves as a prohormone for two other steroids—*dihydrotestosterone* and

dihydrotestosterone

oestradiol—that are active in target tissues (see Figure 20). Certainly the fetus demonstrates this conversion, and dihydrotestosterone is responsible for the *masculinization* of the external genitalia and male urethra. When female embryos are treated with testosterone, they develop *both* male and female gonads and ducts.

masculinization*

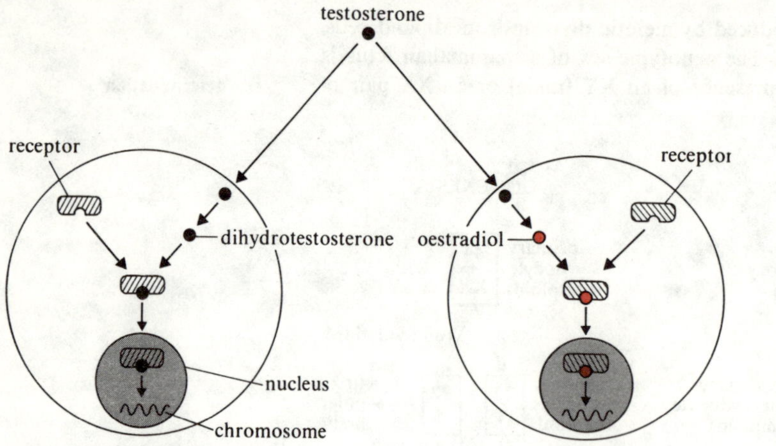

FIGURE 20 Testosterone can be converted into oestradiol or dihydrotestosterone in different target cells in the hypothalamus. These products then bind to cytoplasmic receptors and subsequently modify gene expression in the nucleus of the hypothalamic cell (see Unit 16).

☐ How can you explain this?

■ By postulating that the testes produce another hormone, which diffuses across to the rudiment of the ovary and suppresses its development. Female embryos lack this hormone, so female embryos will develop female gonads even when exposed to testosterone.

The other major physiological difference between the sexes, apart from steroid production, lies elsewhere.

☐ Where?

■ In the hypothalamus. Whereas in females, FSH/LH-RF promotes both the constant release and the surges in gonadotropin release, in males, the levels of FSH and LH stay fairly constant.

What then determines the 'sex' of the hypothalamus? Much of the work on this has been done on rats because they are still relatively immature at birth. Neurons within the rat hypothalamus can be distinguished at around day 15 of gestation. By day 17, FSH/LH-RF is found, and LH and FSH are detectable in the fetal pituitary just before birth (day 20). However, the all important blood circulation between the hypothalamus and pituitary does not develop until the first post-natal week. If androgen is injected into new-born (*neonatal*) female rats, this results in sterility and there is no ovulation (see Figure 21). The timing of this injection is critical; it will only work between a few days before birth until 5 days afterwards. Before this time, the pituitary is 'plastic' and releases gonadotropin according to the character of the hypothalamus. In the same way, if an *anti-androgen* is given to a neonatal male rat, the normal masculinization of the hypothalamus is prevented.

neonate, neonatal

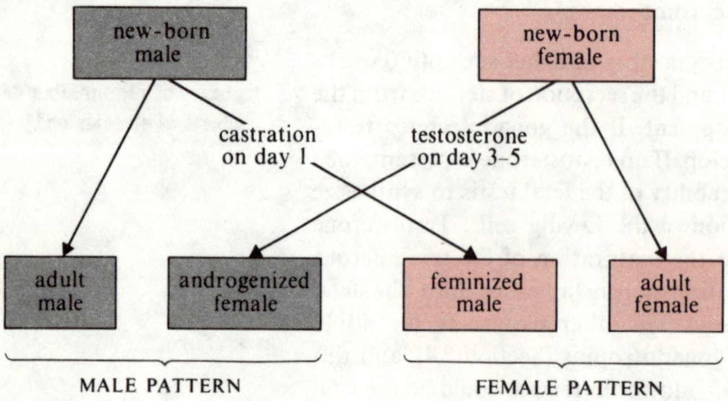

FIGURE 21 Sexual differentiation of the rat brain takes place after birth. In new-born males, testosterone secreted by the testes is converted by target cells in the brain into ŏestradiol, which gives rise to a permanent male pattern of brain structure. If the male is castrated at birth, however, the sexual differentiation of nerve circuits in the brain fails to take place and the brain retains a female pattern. Administration of testosterone to a new-born female rat evokes a male pattern of nerve circuits as a result of the testosterone being converted intracellularly into oestradiol.

In the male, testosterone appears to be responsible for abolishing the potential for cyclical LH release (typical of females); the absence of testosterone in young females results in the hypothalamus developing the capacity to release LH in a cyclical fashion.

Evidence that the effect of testosterone is actually on the hypothalamus itself rather than on the pituitary stems from the discovery that female rats in which the pituitary has been replaced by one from a male will still ovulate. The pituitary is 'driven' by the 'sex' of the hypothalamus!

26

This short account of mammalian sex determination is, of necessity, superficial, and much of it based on the rat. The process appears to be more complicated in other mammals, for example primates. As a result of differentiation, the brain acquires certain permanent characteristics, which underlie, in some way, the sex differences in behavioural responses and neuroendocrine function. Steroids 'mould' the hypothalamus around birth, but this is not the end of the process. Reproductive competence in the rat does not appear for another 6 weeks, when steroids are produced in large quantities; these then trigger the development of secondary sexual characteristics and the changes in behaviour and physiology that are related to reproduction.

Summary of Section 4

Sex is genetically determined (XX = female, XY = male), but on top of this, there are effects during development of steroid hormones. The actions of sex steroids are diverse. Steroids not only maintain the gonads and promote development of the gametes, but they also bring about gonadotropin release via feedback systems. In the developing mammalian fetus, sex steroids are important at three particular stages, summarized in Table 2.

TABLE 2 The effects of hormones on reproductive development in male and female mammals

Stage of development	Male	Female
Early in fetal life	*Testosterone* gives male primary sexual characteristics.	Female primary sexual characteristics and ovaries develop if *testosterone* and *local factors* are absent.
	Local factor from testes suppresses ovarian development.	
At around birth	*Testosterone* makes hypothalamus male suppressing mechanism for cyclical *LH* release.	Mechanism for cyclical *LH* release develops in absence of *testosterone*.
At puberty	Increase in *FSH* and *LH* secretion leads to growth of testes and *testosterone* output.	Increase in *FSH* and *LH* secretion leads to growth of ovaries and higher *oestrogen* output.
	Testosterone gives male secondary sexual characteristics.	*Oestrogen* gives female secondary sexual characteristics.
		Cyclical LH release leads to ovulation.

Objective and SAQ for Section 4

Now that you have completed this Section, you should be able to:

★ distinguish between three phases of mammalian development in which circulating steroids exert key determining effects.

To test your understanding of this Section, try the following SAQ.

SAQ 12 (*Objective 10*)   Predict what would happen if a single dose of testosterone is given to (a) a genetically male rat early in fetal life, (b) a genetically female rat just after it had been born, and (c) a genetically female rat after it had become sexually mature.

5 Contraception

The challenge confronting reproductive research today is a world inhabited by 4000 million people. Each day there is a *net* population gain of 200 000 making a total gain of 70 million every year. If this increase continues at the same rate, then disaster is inevitable. Each year some 50 million abortions are performed, in some cases placing women at risk from surgical infection or causing infertility.

The ideal method of *contraception* should be effective, safe, simple, cheap, reversible and therefore readily acceptable. Several methods approach this ideal, but

contraception*

none, as yet, achieves it. Contraceptive methods have departed in the past three decades from the relatively simple 'barrier' methods refined over the centuries, to the use of steroid hormones encapsulated in the 'pill'. Hailed as a great advance in contraceptive technology in the 1960s, the pill has acquired notoriety because of side-effects associated with its use, and understandably it is now (1980) less popular. In this last Section, we look at the advances in methods of contraception (and their physiological implications) and attempt to point the way to safer and more effective formulations.

steroid contraceptive pill (the pill)*

In its broadest sense, contraception can imply a range of interventions: prevention of gamete maturation, controlled ovulation, prevention of fertilization, prevention of implantation and, most extreme, the termination of implantation or abortion. Advances have been made in all these areas, but particularly so in the first two.

5.1 Steroids

Steroids exert both a positive and a negative feedback action on gonadotropin release from the pituitary. In the luteal (late) phase of the cycle, oestrogen and progesterone, in combination, inhibit LH and FSH release and prevent follicular maturation. This combination of steroids is the basis for the '*combined-type*' *pill*. As progesterone itself is not adequately absorbed from the gastro-intestinal tract, a synthetic *progestogen* is used. Most available preparations are standardized on a cycle length of 28 days being either 21 days on pills and 7 days off, or 21 days on pills plus 7 days on a 'placebo' (usually in the form of lactose) so that a tablet can be taken every day without interruption. Initially, high levels of steroids were used, but lower doses of steroids are very effective, and new progestogens have been introduced. Most combined-type preparations are now very low in oestrogen.

combined-type pill*

progestogen*

The main site of oestrogen–progestogen action is on the hypothalamus, depressing the release of FSH/LH-RF (by negative feedback) and consequently limiting the secretion of LH and FSH. Feedback inhibition may also affect the pituitary directly. FSH and LH are present in only small amounts in a woman taking the steroid pill, and the ovulatory surge of LH is abolished. Not only is ovulation prevented, but the production of oestrogens by the ovaries is suppressed. However, there are other significant changes in the ovaries and uterus as a result of steroid application. The endometrium of the uterus undergoes change, and menstruation is determined by the length of the course of tablets and not the hypothalamo-pituitary rhythm. Menstruation is shorter (in time) and lighter. The ovaries, when examined, appear to be arrested at an early stage of follicular development, and there is no evidence of fresh corpora lutea being present.

Because of the side-effects of oestrogens, a '*mini-pill*' has been designed, which contains only progestogen. A small amount is taken every single day continuously, and menstruation in about 40 per cent of women settles down to the approximate periodicity of the normal menstrual cycle. The progestogen continuously stimulates the endometrium, but at intervals it seems that the dose is not sufficient to support further growth of the tissue, which then breaks down as a menstrual flow. In contrast to the combined pill, progestogen does not usually inhibit ovulation. Progestogen interferes with meiosis in the ovary, changes oviduct mobility, and will prevent implantation if fertilization should occur. Oestrogen levels remain close to normal, because FSH secretion is not inhibited by progesterone.

mini-pill*

Development of a 'one-off' pill has resulted in the so called '*morning-after*' *pill*. This can be *either* progestogen *or* oestrogen, and large doses are involved. The large dose of progestogen taken just after coitus stimulates the growth of the endometrium, which then breaks down as levels decrease, preventing implantation. Similarly, a series of high doses of oestrogen administered after fertilization will prevent implantation. At present, the efficacy in preventing conception by this method is not high, and large doses of oestrogen can produce nausea.

morning-after pill

Instead of daily doses of steroids, oestrogen and progestogen can be taken either together or separately once a month. The regime involves oral administration on about day 22–25 of each cycle, and the oestrogen is stored in the body fat and slowly released, thus preventing ovulation. Injections and implants of steroids have also been tried. Capsules containing progesterone can be implanted just under the skin, and these are effective for periods of a year or more. Intra-uterine

and intra-vaginal devices can also be used as steroid 'implants' to release progestogen slowly and inhibit ovulation.

While the combined type (oestrogen–progestogen) pill is, at the moment, the most
effective contraceptive, it is by no means ideal. Artificial intervention in any
physiological control system brings with it the possibility of unwanted side-effects.
The most serious risks are associated with the blood circulatory system and are
related to the steroid dose, age of the taker, whether they smoke cigarettes, and
the length of time on the steroid pill. In a recent report from the Royal Colleges of
General Practitioners and of Obstetricians and Gynaecologists, it is recommended
that women on the pill should not smoke, that the pill should not be in use
continuously for more than five years, and that after the age of 35 women should
look to other forms of contraception.

When taking oral contraceptives, the rate of blood clotting can increase, and this
means that there is a greater risk of a clot in a blood vessel. (Normal pregnancy
also increases the dangers of blood clotting, and venous thrombosis is quite a
common complication of an otherwise normal pregnancy.) This danger has
promoted the development of a pill with lower oestrogen levels, and it is probably
safer now to take a low-oestrogen pill for 30 years than to complete one pregnancy! Long periods on the pill can lead to *amenorrhoea* (lack of menstruation)   **amenorrhea**
caused by over-suppression of ovarian activity. The primordial follicles do not
secrete even the small amounts of oestrogen required to restart cyclic activity, but
a very weak application of oestrogen will prime LH release by a positive-feedback
response. Steroids can be used to reduce ovulation and to prevent implantation,
but is it possible to use steroids to prevent sperm production? The answer, of
course, is yes. Androgens, oestrogens and progestogens reduce FSH and LH
secretion in males via strong feedback inhibitory effects on the hypothalamus and
pituitary. In turn, spermatogenesis is affected. Another possibility is the use of
anti-androgens, which suppress spermatogenesis without any irreversible effects.
The search for a useful male contraceptive continues. At the moment, nothing is
as effective as *vasectomy*, a simple and relatively safe operation which involves   **vasectomy**
cutting and then tying the ends of the vas deferens. Vasectomy is, however, a
sterilization technique and is not necessarily reversible.

5.2 Novel peptides

Steroids can inhibit the release of gonadotropins either by acting on the hypothalamus (reducing FSH/LH-RF) or by direct action on the pituitary. There is
another way of suppressing pituitary gonadotropin release—by the use of
analogues of FSH/LH-RF.   **peptide contraceptive**

☐ How would this work?

■ The analogue could be an *antagonist* that would bind to the FSH/LH-RF
receptor in the pituitary but would not activate LH and FSH release from the
pituitary.

With the isolation and identification of the peptide FSH/LH-RF, both Roger
Guillemin and Andrew Schally (Unit 16, Section 7) were quick to realize the
possible application of their work. The structure of FSH/LH-RF is a sequence of
10 amino acids:

glutamate–histidine–tryptophan–serine–tyrosine–glycine–leucine–arginine–proline–glycine–NH$_2$

    1   2   3   4   5   6   7   8   9   10

This peptide has been synthesized and appears to be active in all vertebrates, that
is, it is not species specific. Between 1972 and 1977, Schally synthesized 300
analogues of this peptide in an attempt to discover which sequences were
important.

Some analogues are superactive agonists, that is, they are more potent in eliciting
prolonged LH and FSH release than the natural peptide. Others are antagonists
to FSH/LH-RF; they bind to the receptors, but do not trigger LH and FSH
release, and therefore also block any naturally circulating FSH/LH-RF activity.
Peptides are naturally broken down by *endopeptidase* enzymes. If different amino   **endopeptidase**
acids are introduced into the peptide, this breakdown may be inhibited and
consequently the potency of the peptide, whether agonist or antagonist, will be
enhanced because it is not so readily destroyed.

If histidine (2) and glycine (6) are replaced with D-phenylalanine, then a potent antagonist is produced that inhibits LH and FSH release for 6–8 hours after injection. The value of inserting a D amino acid is that these amino acids are particularly resistant to enzyme attack (the L form of an amino acid is the naturally occurring form).

Such antagonists will also suppress LH and FSH release in males. The implications of this work are enormous! A small peptide, cheaply synthesized and with long lasting, but reversible, effects, could be the new and safe method of contraception. It could be used in men *and* women, probably with few, if any, side-effects. While more work is needed on the effects of a long-lasting block of FSH/LH-RF receptors, and on the timing and administration, it may represent a significant breakthrough.

☐ What may a major problem of the administration of a peptide be?

■ Peptides cannot be given orally in the form of a pill because digestion would inactivate them. The only serious complication, then, might be the route of administration: injection or long term implants might be necessary.

Another new approach to contraception is to exploit the fact that receptors will become less sensitive with prolonged exposure to their chemical signals (receptor down-regulation, see Unit 16, Section 8). If FSH/LH-RF is repeatedly given to a woman, ovulation will cease because the pituitary receptors to this releasing factor lose their sensitivity to it. Small amounts of a superactive agonist of FSH/LH-RF can therefore be used to block ovulation. Similarly, such superactive releasing factors will block sperm production in males. In both cases the down-regulation is easily reversible, an important criterion for any contraceptive.

Summary of Section 5

1   The contraceptive pill is a mixture of an oestrogen and a progestogen, or progestogen alone. Steroids taken daily suppress the release of FSH/LH-RF from the hypothalamus (and pituitary release of FSH and LH) by a negative feedback action. These steroids therefore prevent follicular maturation and ovulation.

2   Progestogen or oestrogen taken in large doses after fertilization prevent implantation.

3   Major and minor side-effects of the steroid contraceptive pill are an increased rate of blood clotting, amenorrhoea and disruption of menstruation.

4   Steroids also affect spermatogenesis via a feedback action on FSH and LH secretion in males.

5   Analogues of the peptide FSH/LH-RF may provide useful alternatives to steroids as contraceptive agents.

Objectives and SAQ for Section 5

Now that you have completed this Section, you should be able to:

★ explain the contraceptive actions of daily doses of steroids.

★ show how analogues of a hypothalamic releasing factor could be used to prevent gametogenesis and ovulation.

To test your understanding of this Section, try the following SAQ.

SAQ 13 (*Objective 11*)   Which of the following treatments should, in theory, have contraceptive effects?

(i) One injection of LH.

(ii) Daily oral administration of a peptide that is anti-LH.

(iii) Injection of progesterone as an implant that slowly releases the hormone.

(iv) Daily oral administration of oestrogen and progestogen.

(v) Daily oral administration of prolactin.

(vi) A single injection of FSH.

(vii) Daily injection of an FSH/LH-RF antagonist.

6 Concluding remarks

In this Unit we have concentrated on the role of hormones and the nervous system in the maturation of sperm and eggs and in the timing of the delivery of sperm *to* eggs. Little attention has been given to the processes of implantation, fetal physiology and parturition.

As indicated earlier, it is not meaningful to generalize from the actions of hormones in a single mammalian species, and while much of the discussion has centred around the human reproductive cycle, on a number of occasions descriptions have been provided of similar events in a range of species. Such comparisons in the Unit are not gratuitous, and you should, by now, have an idea of the considerable variation between mammals in terms of oestrogens, androgens and peptides from the hypothalamus and pituitary and their role in timing and inhibiting the production of gametes.

It is perhaps not surprising that we, as a species, should be so intrigued with our own reproductive physiology. The menstrual cycle is unusual though. Most mammals are seasonal, coming into breeding condition at a particular time of year in response to changes in light, temperature or rainfall. The result is ovulation and conception, and after a period of gestation, parturition occurs at a time that is propitious for the survival of the offspring. Should the first ovulation not result in fertilization, then the female will undergo a cycle of events that again results in ovulation. These cycles will repeat during the breeding season until fertilization is successful or until the season ends.

In one sense, therefore, the reproductive *cycle* is an unusual event and can be thought of as a fail-safe mechanism. In women, menstrual cycles occur over a period of 30–40 years, broken only by pregnancy or drastically changed by the ingestion of steroids. This 'uncoupling' of reproductive readiness from environmental cues and from other 'natural' contraceptives (such as that provided by suckling) has necessitated the tremendous efforts that are being made towards finding an acceptable form of contraception. An understanding of how natural methods of fertility operate in other species may provide us with the answer.

Objectives for Unit 18

1 Define, recognize or place in the right context all the items marked with an asterisk in Table A.

2 Describe, in words and diagrams, the basic anatomy of the male and female reproductive systems. (*SAQs 1 and 2*).

3 List the sequence of events in an oestrous cycle, and describe the differences between oestrous cycles typified by: apes and humans (menstruators); ferrets, cats and rabbits (induced ovulators); and rats. (*SAQ 1*).

4 Describe the major endocrine changes that control the menstrual cycle. (*SAQs 3 and 4*).

5 Provide evidence for a hypothalamic–pituitary–gonadal link in the control of the menstrual cycle. (*SAQs 5, 7 and 8*).

6 Distinguish between negative and positive feedback control of gonadotropin release. (*SAQs 3 and 8*).

7 Describe the functions of the corpus luteum and distinguish between luteotropic and luteolytic factors. (*SAQ 6*).

8 Give two examples where the nervous system plays a role in the timing of reproduction. (*SAQs 9, 10 and 11*).

9 Discuss the evidence for external influences on the timing of reproduction. (*SAQs 9, 10 and 11*).

10 Distinguish between three phases of mammalian development in which circulating steroids have key regulating effects. (*SAQ 12*).

11 Describe how steroids and peptides can be used as contraceptive agents. (*SAQ 13*).

SAQ answers and comments

SAQ 1 The correct statements are (iv), (v) and (vii).

(i) Ovulation involves the release of the ovum *from* a mature follicle. It is then picked up by the funnel of the Fallopian tube, which lies close to but not attached to the ovary, and is carried down into the uterus. (Section 1.1).

(ii) The corpus luteum is a transformed ruptured follicle that persists in the *ovary* after ovulation. If pregnancy occurs, it remains in the ovary and secretes a steroid hormone important for the maintenance of the uterine wall. (Section 1.1).

(iii) The gestation period is the time from implantation until birth. (Section 1.1).

(vi) Most mammalian species are seasonal breeders, and the gonads regress out of season. (Section 1.2).

SAQ 2 Blockage of the Fallopian tubes may prevent descent of the ovum to the uterus but does not necessarily prevent sperm reaching the ovum in the tube. Fertilization can occur, and sometimes the blastocyst will implant in the tube itself. An ectopic pregnancy such as this (or in extreme cases when fertilization takes place before the ovum enters the tube) is dangerous because the fetus being in the wrong position leads to local vascularization of tissues.

SAQ 3 (a) This is the peak concentration of progesterone, and you would expect it to inhibit LH secretion (and therefore ovulation) and possibly reduce the effect of FSH on the follicles. The endometrium will be maintained in a 'progestational' state, and there will be no menstruation. The high level of progesterone (because of the lack of LH and thus corpus luteum formation) will depress natural progesterone levels.

(b) Low levels of progesterone would not affect the ovaries or the endometrium, and circulating levels of steroids should not be affected.

SAQ 4 (a) Testosterone alone will probably suffice to maintain the seminiferous tubules, allow spermatogenesis and maintain the other glands and ducts for this period. LH might be safer because it may act directly on meiosis as well as causing testosterone release.

(b) FSH is needed for the Sertoli cells, and once again it is probably better to give LH than testosterone in the long term.

SAQ 5 (ii) is the expected result. LH stimulates the Leydig cells to secrete more androgen so (iii) and (i) are wrong. Spermatogenesis is under FSH control so (v) is unlikely. (iv) is clearly wrong because there is little evidence for cycles of sexual activity in the human male.

SAQ 6 The accurate statements are (ii) for (a) rats and (v) for (b) humans.

(i) is wrong; the implanted egg does not secrete a hormone.

(iii) is wrong; after implantation the corpus luteum is retained.

(iv) is wrong; prostaglandins are thought to be luteolytic and will cause abortion.

(vi) is wrong; the corpus luteum produces progesterone, which maintains the endometrium of the uterus.

SAQ 7 Your flow diagram should look something like Figure 22.

SAQ 8 Your flow diagram should look something like Figure 23.

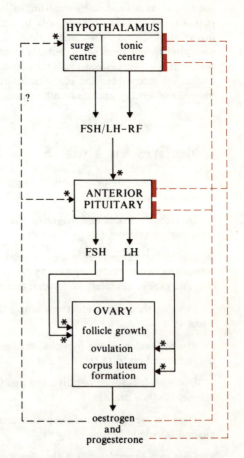

FIGURE 23 The products of the ovary (oestrogen and progesterone) have different effects at different times of the menstrual cycle. Look at the right hand side of the figure first. This depicts the negative feedback effects of oestrogen and progesterone on the pituitary and hypothalamus that are operating during the follicular and luteal phases of the cycle. The left hand side of the figure illustrates the positive feedback effect of oestrogen that is operating just before, and in fact is responsible for, ovulation. Note that different parts of the hypothalamus release FSH/LH-RF and respond differently to steroids.

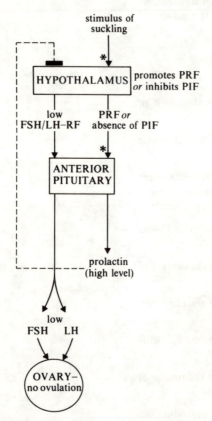

FIGURE 22 Suckling stimulates the production of prolactin from the anterior pituitary either by promoting the release of PRF or inhibiting the release of PIF from the hypothalamus. Prolactin in the blood feeds back onto the hypothalamus and inhibits the release of FSH/LH-RF. The consequence is low levels of the gonadotropins FSH and LH. Therefore the follicles in the ovary do not mature, and so ovulation cannot take place.

32

SAQ 9 As the doe is an induced ovulator, the rabbit cannot have ovulated before hypophysectomy. In the case of the rat, it will depend entirely on the stage of the oestrous cycle. If in oestrus, an injection of LH will have the same effect as in the rabbit. If the rat had just ovulated, then nothing would be necessary to allow fertilization.

SAQ 10 Examine blood or urine for signs of cyclical steroid changes; see whether fertility is affected by anaesthesia of the genitalia or cervix or by cutting the sensory nerves from the pelvic region before mating. The ovary of an induced ovulator is affected by two separate classes of environmental effect—stimulation from the male (which induces LH release) and the normal influences of light acting on the hypothalamus, which affects the onset of reproduction in both spontaneous and induced ovulators.

SAQ 11 A pheromone could act through the olfactory system and the hypothalamus to advance the cycles of those ewes not yet in oestrus. Alternatively, a visual or auditory stimulus could have the same effect.

SAQ 12 (a) A genetically male rat will start to secrete testosterone as the gonads develop. Giving testosterone may simply speed up the developmental process or have no effect.

(b) A genetically female rat at birth has ovaries and will be producing oestrogens. An injection of testosterone at this time causes sterility, because it will abolish the ability of the hypothalamus to release LH from the 'surge' centre. An androgenized female does not ovulate.

(c) Giving testosterone to a sexually mature female would probably prevent the release of gonadotropins from the pituitary by a negative feedback system and therefore produce a temporary halt to follicular maturation and ovulation. This would not be a permanent effect, unlike (b).

SAQ 13 The correct answers are (iii), (iv), and (vii).

(i) LH injection would initiate ovulation.

(ii) Oral administration of a peptide would not work because it is liable to digestion in the gut.

(iii) Progesterone at a high level through the cycle would prevent release of FSH and LH and therefore prevent follicular maturation and ovulation.

(iv) This is the common formula of the contraceptive 'pill'.

(v) Prolactin is a peptide and oral administration would render it useless.

(vi) FSH injection would initiate follicular maturation.

(vii) If the hypothalamic control of pituitary gonadotropin release is blocked, follicular maturation and ovulation will be prevented.

Further reading

Holmes, R. L. and Fox, C. A. (1977) *Control of Human Reproduction*, Academic Press, London.

Slater, P. J. B. (1978) *Sex Hormones and Behaviour*, Studies in Biology Series, Arnold London.

Johnson, M. and Everitt, B. J. (1980) *Essential Reproduction*, Blackwell, Oxford.

Acknowledgements

Grateful acknowledgement is made to the following for permission to reproduce material in this Unit.

Figures 2 and 6a M. H. Johnson and B. J. Everitt, *Essential Reproduction*, Blackwell, 1980; *Figure 9* P. J. B. Slater, *Sex Hormones and Behaviour*, Institute of Biology's Study in Biology no. 103, Edward Arnold, 1978; *Figures 15 and 16* Photoperiodism and Seasonal Breeding in Birds and Mammals, B. K. Follett, in *Control of Ovulation* (Crighton et al., eds), Butterworth, 1978; *Figures 20 and 21* Behavioral Sequelae of Prenatal Hormonal Exposure in Animals and Man, A. A. Ehrhardt, in *Psychopharmacology: A Generation of Progress* (M. A. Lipton et al., eds), Raven Press, 1978.

List of Units

The Diversity of Organisms

Cell Biology

Development

Animal Physiology

Plant Physiology